Microemulsions
THEORY AND PRACTICE

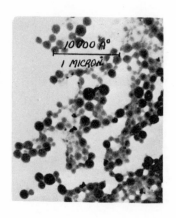 

I   II

(I, II) Electron micrographs of alkyd-in-water microemulsions: (I) average diameter of droplets 1200Å. Magnification 17,000 X taken at 10,000 X; (II) average diameter of droplets 300Å. Magnification 140,000 X taken at 80,000 X. (III) Phase equilibria diagram representing four component micellar solutions of Friberg school. (I and II reproduced from Prince, L. M., "Carnauba Wax Molecules," Soap and Chemical Specialities, September, October, 1960, courtesy MacNair, Dorland, Inc.)

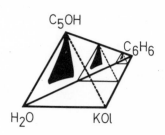

III

Microemulsions
THEORY AND PRACTICE

Edited by

Leon M. Prince

Consulting Surface Chemist
Westfield, New Jersey

ACADEMIC PRESS, INC.   New York   San Francisco   London   1977

A Subsidiary of Harcourt Brace Jovanovich, Publishers

Academic Press Rapid Manuscript Reproduction

ACADEMIC PRESS, INC.
111 Fifth Avenue, New York, New York 10003

United Kingdom Edition published by
ACADEMIC PRESS, INC. (LONDON) LTD.
24/28 Oval Road, London NW1

Library of Congress Cataloging in Publication Data

Main entry under title:

Microemulsions.

    Includes bibliographical references and index.
    1.   Emulsions.   I.   Prince, Leon M.
TP156.E6M52   660.2'84292   77-5362
ISBN 0–12–565750–1

To my wife Adelaide, without whose inspiration, patience, and forbearance this book would not have been possible; and we are all grateful to my daughter Judith, who assumed the responsibiity of coordinating the completion of this book during my illness.

Contents

List of Contributors

Vinod K. Bansal
Departments of Chemical Engineering and Anesthesiology
University of Florida
Gainesville, Florida 32611

Stig Friberg
Department of Chemistry
University of Missouri-Rolla
Rolla, Missouri 65401
and the Swedish Institute for Surface Chemistry
Stockholm, Sweden

Hironobu Kunieda
Department of Chemistry
Faculty of Engineering
Yokohama National University
Yokohama, Japan

Leon M. Prince
Consulting Surface Chemist
7 Plymouth Road
Westfield, New Jersey 07090

Dinesh O. Shah
Departments of Chemical Engineering and Anesthesiology
University of Florida
Gainesville, Florida 32611

Kozo Shinoda
Department of Chemistry
Faculty of Engineering
Yokohama National University
Yokohama, Japan

Preface

Stable, opalescent "emulsions" had been on the market for thirty years when the term "microemulsions" was coined in 1958. "Microemulsions" describe the systems that Hoar and Schulman scientifically identified in 1943 as special colloidal dispersions. With this perspective, the book seeks to deal with both theory and practice regarding microemulsions so that the make-up of these unusual systems will be understandable to the practical chemist and the theoretician.

During the last five years, interest in microemulsions has expanded enormously, spawning a multitide of papers on every aspect of the subject. This circumstance has prompted the present book, which is the first text to be completely devoted to microemulsions, presenting both a history of the development of industrial emulsions and a sophisticated account of emulsion theory.

This book is at a crossroads in the development of the theory of these colloidal systems. During the past several years, there have been proposals that Schulman's fluid, translucent, isotropic dispersions of oil and water might not be emulsions at all but instead are swollen micelles, micellar solutions, or micellar emulsions. Such systems have been quantitatively defined by means of phase diagrams. This difference in terminology appears to be more than just a question of semantics and this book presents both sides of the controversy. The case for microemulsions vs micellar solutions was presented by Shinoda and Friberg in 1975 under the title "Microemulsions: Colloidal Aspects." The critical evaluation of the investigation of microemulsions stabilized by mixed films has certain areas of conflict with this review, which will require critical attention.

In addition to discussing theory, this book also sets forth the physics and chemistry of these systems in practical terms. It teaches a basic understanding of microemulsions, enabling the reader to follow the adsorption of molecular species to the interface between oil and water and to comprehend the interactions between them that determine the direction and degree of curvature of the interface and thus the type and droplet size of the resulting dispersions of liquids. HLB is only discussed briefly, since the reader is assumed to be familiar with this rating system for emulsifiers and oils. Although the references contained herein are comprehensive, covering most of the field, they are not complete. Many discussions of the theory of microemulsions have appeared in articles that deal with unrelated subjects. Finally, where primary references have been inadvertently omitted, they may usually be found in the papers to which such reference has been made.

Micellar solutions, which are usually described in terms of phase equilibria diagrams, are discussed, but again a primer course is not given. Enough basic material is presented to enable the reader to obtain a grasp of the subject.

Since the writing of this book certain aspects of the subject have undergone important changes. It has come to be understood that the phase equilibria diagram defines a very specific, narrow, and critical range of systems and requires much tedious and precise work to make. The use of phase maps is a much simpler device, which includes a broader range of experimental data that are much easier to accumulate. It is anticipated that future investigators will take the simpler route.

Within the past several months, there has been a realization on the part of the editor that some of the data on this subject might have been based on uncertain premises. Though a great deal is known about microemulsions, certainly we are just beginning to fully understand this phenomenon. The editor is reminded of the words of Gilbert N. Lewis:

> Science has its cathedrals built by a few architects and many workers . . . . The scientist builds slowly and with a gross but solid kind of masonry. If dissatisfied with any of his work, even if it be near the very foundation, he can replace that part without damage to the remainder.

Schulman's Microemulsions

LEON M. PRINCE

Consulting Surface Chemist
7 Plymouth Road
Westfield, New Jersey   07090

I. INTRODUCTION

Three kinds of dispersions of oil, water, and surfactant will be considered in this book: microemulsions, micellar solutions, and the common variety of emulsions, which will be called macroemulsions. The last variety appears white and is characterized by its propensity to separate into its two original liquid phases on standing. Microemulsions and micellar solutions do not appear white; they are translucent or transparent and do not separate. Neither of the latter has been characterized well enough at this point to make possible a clear-cut differentiation between them. The writer has chosen to identify microemulsions as having droplet diameters in the range of 2000 A (0.2 μm) down to 100 A (0.01 μm) and micellar solutions as having aggregate diameters less than 100 A. In a sense, this is an extension of Becher's definition (1) of an emulsion as an unstable heterogeneous system in which the diameters of the dispersed droplets, in general, exceed 1000 A. This writer's definition is by no means a firm characterization and is open to legitimate dispute. It is mentioned here simply to give the reader a rough idea of, and the general range of, the area in question.

Clayton (2) had no such problem. His definition unmistakably conveys the comprehensive insight into these systems that workers in this field had 30 years ago. In fact, Clayton graciously acknowledged that F. Selmi quite clearly gave this same definition almost 100 years earlier (3). Clayton's definition was "An emulsion is a system containing two _liquid_ phases, one of which is dispersed as globules in the other. That liquid which is broken up into globules is termed the dispersed phase, whilst the liquid surrounding the globules is known as the continuous phase or dispersing medium. The two liquids, which must be immiscible or nearly so, are frequently referred to as the internal and external phases, respectively." He further stated that when one of the liquids is water and the other a water-insoluble liquid or "oil," two sets of emulsions are theoretically possible, depending upon whether oil is dispersed in water (o/w) or vice versa (w/o). Also, he considered that there is no _a priori_ reason why emulsions of any desired concentration should not be made of either either liquid in the other.

In addition, there may be mixed emulsions in which both o/w and w/o emulsions are present in the same system. The oils may be different. One example is a margarine in which a highly polyunsaturated oil-in-water emulsion of a very small droplet size (0.5 to 1 μm) is the internal phase of a normal w/o margarine (4), the dispersed droplets of water of which are in the 20-25 μm range. This is called a double emulsion. There may be more complicated systems.

In more modern terminology, an emulsion may be defined as a dispersion of two (or more) mutually insoluble liquids, one in the other. Because of the surface tension forces at play between the two liquids, the dispersed phase consists of spherical droplets. The state of the art at the time Clayton promulgated his definition contemplated droplets only in the diameter range of a few microns to a hundred microns. The droplets of Schulman's microemulsions were very much smaller than these and were, therefore, at first thought to be micelles because of the way they scattered light and their stability. In the case of microemulsions and micellar solutions one of the liquids is always water; in the case of macroemulsions, water is usually one of the liquids but there are macroemulsions in which neither liquid is water (5).

Clayton's definition of an emulsion envisioned a system of

two mutually insoluble liquids, one dispersed in the other,
(a) without any emulsifier, (b) with an emulsifier consisting
of a single layer of finely divided powders situated at the
interface between the two liquid phases, or (c) a single layer
of discrete molecules adsorbed in highly oriented form at the
interface. This oriented single layer is variously called a
monolayer, a monomolecular film or layer, an interface or an
interphase, the latter term suggesting that the single layer
of highly oriented molecules constitutes a separate, third
phase. All of Clayton's systems have a common denominator:
mechanical work in the form of agitation, homogenization,
ultrasonication, etc., must be put into the system in order
to disperse one liquid in the other in the form of droplets.

The concept of a mixed film was part of (c) above. This
was a monomolecular film consisting of two species, a soap or
detergent and an alcohol (cosurfactant) which interacted with
each other in the film in such a way as to greatly increase
the stabilizing power of the interface by reducing droplet
size. Input of work, however, was still required to form
these systems. Schulman called the alcohol an amphiphile,
a term that has gone into disrepute in recent years.

In the very year that Clayton's last edition appeared,
1943, Hoar and Schulman introduced a fourth situation. This
was a mixed film which, in combination with certain oils, was
capable of generating an emulsion of very small droplet sizes
without the input of any work, i.e., the emulsion formed spontaneously. Such systems came to be called microemulsions.

Like macroemulsions, microemulsions are of the water-in-
oil (w/o) type and oil-in-water (o/w) type and invert from one
type to the other by adding more of one phase or by changing
the type of emulsifier. It is in this area of inversion that
Schulman's microemulsions display a peculiar phenomenon.
Beginning with a fluid w/o microemulsion, as water is added,
they pass through a <u>viscoelastic gel stage</u>. As more water is
added, they invert to an o/w emulsion which is fluid again.
This process is reversible. It has been determined that the
viscoelastic gel stage (which can be almost solid) is comprised of an hexagonal array of water cylinders adjacent to
the w/o stage and a lamellar phase of swollen bimolecular
leaflets adjacent to the o/w emulsion. The phases of the gel
stage are called liquid crystalline phases and this perhaps
more than any other single factor is responsible for the controversy of microemulsion vs. micelle as discussed in Chapters 5 and 6.

It is relatively easy to recognize a commercial microemulsion of the Schulman type. Providing you know that oil,
water, and surfactant are in the system, the chances are very
good that the dispersion is a microemulsion if it is fluid,

optically clear (transparent) or opalescent (translucent) AND
when spun in a laboratory centrifuge for 5 minutes at 100 G's,
the dispersion does not separate. The reasons to support this
empirical definition are presented in some detail below. As
will be noted, these are essentially physical in nature. The
chemistry of the microemulsion upon which the formulation of
these extraordinary systems are critically dependent, is
treated in the third chapter.

 This chapter enumerates many of the microemulsions of commerce and describes methods for identifying them. The treatment is as practical as possible consistent with the scientific principles involved.

II. MICROEMULSIONS OF COMMERCE

 Microemulsions are quite ubiquitous. We encounter them
frequently in our daily lives. When we arise in the morning
we find them as transparent essential oil-in-water (o/w) types
in our mouthwash or men's shaving lotion. At breakfast, homogenized milk contains microdroplets of fat dispersed in water,
and nonfat dried milk powers instantly form microemulsions
when stirred into water. Our clothing is returned from the
dry cleaners having been cleaned with transparent microemulsions of water dispersed in a dry cleaning fluid. There is a
good likelihood that the fat we eat for breakfast is absorbed
in our intestines via a microemulsion or micellar process. If
we wax the kitchen or family room floor, it is probably done
with an emulsion polymer latex that can be classified as a
microemulsion. Deodorizers and sanitizers to clean floors and
porcelain may be solubilized--another way of describing a
clear, transparent microemulsion. Flavors in pre-mix cake ic-
ings, ice cream, gelatin, desserts, as well as the flavors in
many beverages and fountain extracts are often solubilized.
Many food products contain or are entirely microemulsions.
Pomades of the "ringing gel" type are a form of microemulsion,
the viscoelastic gel stage. Cold creams and other cosmetic
products may be microemulsions. Microemulsions are used to
uniformly disperse the active principles of pharmaceutical
preparations (e.g., vitamins). Many industrial cleaners in
the janitor's supply trade are microemulsions. In recent
years slugs of w/o microemulsions have been used to recover
more oil from old wells (tertiary oil recovery). This is an
impressive list and one which is destined to become longer as
we learn how to microemulsify oils of different chemical
composition.

 For many years prior to 1959, before the term microemulsion was coined, colloidal dispersions of this kind were
called emulsions but they occupied a special niche in the

marketplace because their stability was measured in years. They were easily recognizable by their transparency or opalescence. Some of these systems were completely transparent but most of them were of the water-in-oil type. Such systems were considered unique because they looked and behaved differently than ordinary macroemulsions, which scattered white light and whose stability was measured in hours, days, or, at best, months. Among the prominent o/w microemulsions of this era were the Carnauba wax floor polishes, pine scrub soaps, cutting oils, chlordane emulsions, alkyd emulsions, all of the o/w type; only the cleaning fluids, which were also prominent, were of the w/o type. The evolution of these products is described in the next chapter. Suffice it to say here that of these pioneering products, only the cutting oils, pine scrub soaps, and cleaning fluids occupy an important place in today's marketplace. However, direct descendants of these products are gaining an ever-growing share of the market.

When Schulman called these systems microemulsions, he put them in their own niche, differentiating them from other classes of liquid/liquid systems. By doing this they could be viewed in a clearer scientific perspective, which enabled a number of advancements to be made in both the theory and art.

All microemulsions, however, cannot be classified as emulsions in accordance with Clayton's broad definition. The usual classification of microemulsions, or for that matter macroemulsions, as dispersions of oil and water is not strictly correct. As Clayton pointed out, an emulsion is a dispersion of any two insoluble liquids, one in the other. The fact that one is usually water is only a practical consideration. Neither liquid need be water. For example, glycerine is insoluble in olive oil, carbon tetrachloride, and amyl acetate, and emulsions have been made of these pairs of liquids (5). Many other mutually insoluble pairs of liquids can be emulsified and, depending on their relative indices of refraction and the emulsifying agent used, they may be transparent or chromatic while their droplets are in the micron size range. None of these systems have as yet made their debut on the market.

On the other hand, when water is one of the liquids, the definition of an oil is subject to a much broader definition than might be expected. For emulsification purposes an oil may be considered as any water-insoluble material that can be made liquid at a temperature at which water remains a liquid. By the use of pressure vessels, this means in practice that solids which melt as high as 220° to 250°F. can readily be emulsified with water. This encompasses high melting waxes, alkyds, polymers, and other amorphous substances as well as some crystalline ones. The range of solids may be increased if they may conveniently be dissolved in a solvent that can be tolerated in the end use. A solid toxicant in deodorized kerosene or xylene as used in agricultural sprays for deposit by plane are commonplace examples of these.

Systems of this kind are called emulsions or microemulsions, depending on droplet size, because they are formed by an emulsification process. Upon cooling, of course, the solid-in-water system is, in the strictest sense, a dispersion. Obviously, the terms wax emulsion, polymer emulsion, or Lindane emulsion are commonplace, and we know what is meant.

As opposed to macroemulsions, microemulsions command a premium in the marketplace. Even though they may not be recognized as such--the transparent ones may be called solubilized systems, micellar solutions, or just solutions--their attributes contribute special value to the product in one or more ways. The most obvious is stability. In the case of wax or polymer emulsions (for floor polish and paint), fine particle size ensures high gloss and film integrity. The opalescence or transparence of cosmetic systems may add an aesthetic touch to their sales appeal. Uniformity of dosage in pharmaceutical preparations is another important contribution of microemulsions. Finally, the rheological properties (cf. C. below) of microemulsions can be adjusted easily so that viscous or fluid systems can be obtained almost at will.

III.   PHYSICAL PROPERTIES

There are two reasons for wanting to know about the physical properties of microemulsions. The first is to be able to recognize such systems when they are met. This involves identification by the naked eye and, at least, a minimum comprehension of the kind of colloidal system one will have to deal with. The second has to do with the ability to understand the performance of the systems and what measures can be taken to modify their performance attributes to your benefit. Some of this material has already been presented in unified form (6).

Among the physical measurements that are useful for identifying microemulsion systems are light scattering, optical birefringence, sedimentation, centrifugation, rheology, conductivity, and very recently, nuclear magnetic resonance (NMR). Each measurement tells us something about the system. Consideration of two or more measurements together goes a long way in firmly classifying these oil and water systems. The several measurements are described below in simplified form. All of these techniques are within the purview of the practical chemist or biologist. More scientific discussions of these techniques are considered in other chapters, where their greater complexity is relevant.

A. Light Scattering

The most obvious property of systems of very small aggregates is the way they scatter light. In aqueous dispersions this is apparent to the naked eye as the Tyndall Effect. It is that portion of the scattered light that is polarized. Thus, microemulsions appear blue to reflected light and orange-red to transmitted light. Dilution of the dispersion accentuates this effect. On the other hand, certain systems will appear transparent in concentrated form and then display opalescence on dilution.

Light is scattered by all molecules or components of them. Particles that are large in comparison with the wavelength of light (white light, λ, averages 5600 A) reflect and refract light in a regular manner and thus appear white, whereas particles that are small in comparison with the incident light waves scatter light in all directions. This scattered light is plane polarized, each particle becoming the source of a new wave front. When the droplets of an emulsion are less than $1/4$ λ (ca. 1400 A) in diameter, white light can pass through the dispersion and it becomes translucent or opalescent. Depending on the relative index of refraction of the oil and water, such systems appear transparent to the naked eye when the droplets are about 100 A in diameter. (Translucent is the general term and includes transparent, which may be called very translucent.)

This phenomenon is not restricted to solutions. Freshly formed cigar smoke will scatter blue light and transmit orange red. Moreover, if one watches a tall exhaust stack spewing out its smoke, the same phenomenon will be observed close to the outlet; after a few feet, however, the particles aggregate into larger ones and then scatter white light.

Lord Rayleigh (J. W. Strutt) ascribed the blue of the sky and the orange red of the sunset to these same phenomena, but the aggregates in his system were molecules of oxygen and nitrogen. Although the ocean of air above us extends almost 1000 miles high, the bulk of the gases exist less than 50 miles from the surface of the earth. Sunlight passes through these gases and is reflected by the surface of the earth. This reflected light is scattered by the 50 mile thick layer of gas. Rayleigh developed an equation that showed that the shortest wavelengths of the mixture that is called white light (λ = 4000-7000 A) are scattered most. Thus, blue light is scattered more than the longer red wavelengths and our eyes see the sky as blue. As the sun lowers in the sky and we are able to look almost directly at it, we see the orange-red hues transmitted by the gases and by the very fine particles of dust close to the earth. We can see only these two basic colors in the sky. Were it not for this phenomenon of nature, we would see a black sky behind the sun as the astronauts did on the moon.

It is noteworthy that it takes 50 miles of scattering to produce the basic colors of the sky when the scatterers (chiefly oxygen and nitrogen molecules) are only an Angstrom unit or two in size. In our beakers and bottles, where the scattering takes place in a few inches, the size of the scatterers must be in the 100 A to 1000 A range for our eyes to see the Tyndall Effect.

Accordingly, the appearance of scattered light may be used to identify microemulsions and to roughly measure the size of their droplets. It is therefore appropriate to tabulate the kind of light scattering the naked eye can see under normal conditions. With the aid of Table 1 and a little practice, the eye can become a very sensitive piece of light-scattering apparatus.

TABLE 1

Visual Guide for Estimating Aggregate Size

Material Structures	Diameter (A)	Appearance to Naked Eye
Water Molecules	2.7	Transparent
Soap Micelles	35-75	Transparent
Micellar Solutions	50-150	Transparent and translucent
Microscopically Resolvable Units	1000-2000	Translucent when dispersed
Macroemulsions	2000-100,000	Opaque, milky
Visually Resolvable Units	500,000	Discrete aggregates

Where more quantitative information is required, as in research studies or quality control, light scattering instruments are available to measure the size of microemulsion droplets with considerable precision. Such instruments are the Brice Phoenix photometer, the Oster microphotometer, etc. With a minimum of calibration they may be used as quality control instruments to ensure that the droplet size of the microemulsion is within certain specifications. This is a particularly useful tool in the case of emulsion polymerization where wide variations in droplet size distribution and average size may be encountered because of the many variables in processing. These instruments also serve as a check on your visual observations of a particular emulsion.

Whereas the relationships between particle size and scattering as presented in Table 1 are essentially correct, these optical effects are, as already mentioned, also dependent upon the index of refraction of the two (or more) liquids and upon the concentration of the droplets. For example, a droplet of a high index of refraction material (e.g., 1.6) like polystyrene in a latex would need to be 80 A in diameter instead of 100 A to appear clear and transparent in water (index of refraction 1.33). Most materials have lower indices of refraction than 1.6, so this introduces only an occasional error. Of more consequence is the fact that one will frequently encounter an o/w microemulsion system which appears transparent at 10-25% oil but will strongly scatter light when diluted to 1 to 0.1% oil concentration. Because of this, dilution is a recommended procedure for detecting microemulsions of this type with the naked eye.

Visual recognition of microemulsions should not be taken lightly. In fact, the microemulsion chemist should train himself carefully in this art. Use of sunlight rather than an artificial source of light is recommended. The eye is better than a microscope because the limit of resolution of a light microscope in blue light is only about 0.1 μm so that droplets smaller than 1/4 λ cannot be seen. However, in a dark-field microscope or ultramicroscope, one sees a sea of scintillating flashes of light which, although very beautiful, give you no clue as to the dimensions of the particles other than that they are less than 1/4 λ (0.14 μm).

This brings us to another aspect of light scattering. The degree and kind of light scattering is also dependent upon droplet size distribution. There are two facets to this.

When droplet size distribution is very wide, i.e., the system is heterodisperse, the scattering by a few large droplets may mask the scattering pattern created by many smaller ones. Homogenized milk is a good example of this. The emulsion looks dead white on casual examination. If, however, one swirls the milk up on the sides of the glass, Tyndall scattering is seen in the thin film. This heterodispersity also affects sedimentation rates. The bombardment of the few larger particles by the many smaller ones slows up the rate of "creaming" beyond the shelf life of the product. Thus, homogenized milk is mechanically and kinetically stable for all practical purposes, although it is not completely a microemulsion and not thermodynamically stable.

It is of interest in this connection that because of the way that most microemulsions are made, as the average droplet size decreases, the distribution becomes more uniform. By the time that the average droplet size is as small as 400 A (0.04 μm), the droplets are quite uniform (cf. frontispiece).

In special cases where droplets or aggregates are extremely uniform in size, colors other than blue or orange-red can be seen when viewed at different angles to the incident beam of light. The writer made a transparent microemulsion of vinyl acetate, water, and Pluronic F-68 which when subsequently emulsion polymerized produced a translucent microemulsion that exhibited a distinct pastel green color superimposed on the usual Tyndall scattering. LaMer (7) noticed similar behavior with sulfur sols of very uniform size and made the following observations: "The purity of spectral colors to the naked eye furnishes a trained observer with a ready qualitative estimate of the degree of monodispersity, while the number of spectral orders gives a ready rough measure of the size. In the size range in which the orders occur, brilliant colors imply a strictly monodisperse character, while pastel shades indicate less monodisperse distributions. Opalescence is a mark of polydispersity." These phenomena are closely related to Mie scattering, which became a practical tool only with the advent of the computer after World War II.

There is one final aspect of light scattering which plays an important role in the Phase Inversion Temperature (PIT) as discussed by Professor Shinoda in Chapter 4. Opalescence occurs over a wide range of temperatures <u>above the critical point</u> of the complete miscibility of liquids. Such opalescence is due to the association of the molecules of one of the components of the system to form aggregates of colloidal dimensions. Above this critical temperature, a formerly transparent microemulsion or micellar solution scatters light that is visible to the naked eye, so that we see the system as translucent or opalescent.

B. Birefringence

Although optical, streaming birefringence is a light scattering phenomenon, it has been set apart here because it requires a source of polarized light and some instrumentation is required to observe it. The apparatus is quite simple, consisting merely of crossed nicols (which may be of the plastic polarized film variety), two microscope slides, and a

source of strong, white light. The appearance of birefringence indicates that the dispersed phase of an oil-water-
surfactant system is no longer in spherical form and that a
liquid crystalline phase is present.

When very small aggregates are not isotropic, i.e., one
dimension is longer than the other, as in rodlike or disclike
aggregates, dispersions of them become doubly refracting when
they are stirred or allowed to stream. Thus, if a drop or two
of an anisotropic dispersion is placed between two microscope
slides and squeezed to produce flow, upon examination between
crossed nicols (which normally extinguish white light completely), the illuminated field will light up into beautifully
colored patterns. This is due to the scattering and repolari-
zation of the polarized light. A particularly striking illus-
tration of birefringence in liquid crystalline phases may be
found in a paper by Wilton and Friberg (8).

This phenomenon is one of the tools used to explain what
happens when a microemulsion system inverts through the viscoelastic gel stage from a w/o to an o/w type. As long as the
aggregates are spherical or isotropic, the system appears
black when viewed through the crossed nicols. However, as
soon as the aggregates turn into hexagonal arrays of water
cylinders or lamellar micelles, they display colored birefringence patterns which do not disappear until the aggregates
change into spheres of oil (or water). This occurs when the
light is polarized. At the same time, without the crossed
nicols, these anisotropic aggregates can appear opalescent to
the naked eye when their size in any dimension exceeds 100 A.
Thus, as a general rule, during the inversion of a clear,
transparent w/o microemulsion, the viscoelastic gel stage will
appear opalescent. In order to keep the viscoelastic gel
stage transparent, a very large ratio of emulsifier to oil is
needed. This also applies to the isotropic o/w microemulsion
stage.

C. Rheology

Microemulsion involvement with rheology occurs in two
areas, both of which were discovered by Schulman (9-11). They
are concerned with the viscoelastic gel stage and the viscosity of the fluid isotropic dispersions.

Optical, streaming birefringence is not the only way to
recognize nonspherical or anisotropic aggregates in systems of
oil, water, and surfactant. A less precise but simpler way is
by their rheological behavior. When dispersed aggregates are
other than spherical they offer more resistance to flow than
their spherical counterparts and this can usually be detected
by a sudden and sharp increase in the consistency or viscosity

of the dispersion. In the case of microemulsions, the formation of the viscoelastic gel stage coincides with the formation of nonspherical aggregates.

Rheology is a branch of Mechanics which in the last few decades has been utilized more frequently to solve everyday technical problems. It deals with the deformation and flow of matter. The paint, printing ink, food, plastic, cosmetic, and pharmaceutical industries, to mention only a few, have found rheology to be a valuable tool in both product development and product control. They have made the terms thixotropy, yield value, dilatancy, pseudo-plastic flow, etc., almost as commonplace as the terms viscosity or consistency.

In the case of the inversion of microemulsions, when the spherical aggregates begin to transfer to cylindrical or lamellar ones, the new forms tend to obstruct the flow of aggregates past one another in the dispersion medium. This produces a high yield value. The dispersion will flow when stress is applied but returns to its original condition when unstressed. In some viscoelastic systems, as in Carnauba wax emulsions, the yield value is so high that temperatures must remain high in order to be able to stir and mix the batch.

From a phenomenological viewpoint, rheological behavior in the form of increased yield values and viscosities is a practical sign of the onset of the viscoelastic gel stage in microemulsion systems. It also tells us that the system is no longer an emulsion but probably is in the form of a liquid crystalline phase. Rheologically it is not possible to tell where the cylinders stop and the lamellar micelles begin.

In the fluid, isotropic systems, viscosity may be controlled by a more subtle phenomenon originating at the molecular level. In their study of macroemulsions and mixed films, Schulman and Cockbain found that the viscosity of their emulsion depended upon the ratio of alcohol to soap or detergent in the mixed film. They found that the higher the ratio of alcohol to soap or detergent, the higher the viscosity of the resulting macroemulsion. Within limits, this trick may be applied to microemulsions. However, since the formation of microemulsions is much more sensitive to the ratio of alcohol to surfactant, the range over which one can control viscosity by this method is restricted. But when one considers that the control of viscosity in a microemulsion is almost impossible to accomplish by means of normal thickening agents, which because their surface activity will destroy the stability of the system, this becomes a very viable tool. How this can be done is described in Chapter 3.

D. Sedimentation

Sedimentation rates measure the stability of emulsions and

accordingly play an important role in differentiating between a macroemulsion and a microemulsion.

In the previous sections, means have been presented for estimating the average droplet size, droplet size distribution and the shape of the dispersed phase of a microemulsion system. This exposition assumed that the microemulsion was already available. But such systems are rare, as the formulator intent on microemulsifying a particular oil can well testify to. The writer knows of only 35 "oils" that are susceptible to microemulsification from a commercial point of view and, further, that the o/w types are much more difficult to obtain than the w/o types. Thus, during the process of formulation a test is required that will provide positive information that a microemulsion has been realized. In combination with light scattering, sedimentation is such a tool.

Sedimentation rates measure the stability of emulsions, differentiating between macro and micro droplets. Normally, if an emulsion breaks on standing, it is recognized as a macroemulsion. But in the formulation of a normal o/w microemulsion droplet sizes are usually in the range of 500 A (0.05 µm) and 200 A (0.2 µm) because of the large amount of emulsifier being used. Such a system will appear opalescent to the eye. In such an average size range, however, it would not be unusual for some of the droplets to be larger in diameter than 3000 to 5000 A, and these in time will separate from the microdroplets. The percentage of such droplets could depend upon processing variables or on the uniformity of the emulsifying agent. In any event, separation of any kind is undesirable. The discussion which follows serves as a guide for identifying satisfactory commercial microemulsions.

Sedimentation velocities may be measured in a gravitational field at 1 G in graduated cylinders in a laboratory centrifuge at 100 to 500 G's or in an ultracentrifuge at extremely high gravitational fields. For practical purposes, the laboratory centrifuge is the quickest way to determine if there are any large droplets in the system. Five minutes of spinning will usually show creaming or sedimentation, depending on the relative densities of the droplets and the dispersion medium. If there is no separation, the chances are good that all the droplets are in the micro range. However, and especially in cases where nonionic emulsifiers are utilized, sedimentation rates at various temperatures should be run. Microemulsions, by their very nature, tend to be temperature sensitive and nonionics are more so, so that testing at temperatures as high as 130 °F. at 1 G should always be a part of the finalizing procedure. Freeze-thaw cycling is a special case and poses a special problem for each system. The ultracentrifuge really finds no place in the formulator's repertoire.

In this connection, however, Schulman told of spinning a water-in-benzene emulsion at 130,000 G's. Some stratification occurred, but upon standing at 1 G for a few minutes the system became homogeneous once again.

The reason for this is called Brownian Movement, after the botanist Robert Brown, who observed in 1827 that small particles suspended in water appeared to be in ceaseless motion. It was not until 1906 that Einstein and Smoluchowski established that the cause of this motion is the bombardment of these small particles by the molecules of water. Particles smaller than 0.5 µm are small enough to absorb kinetic energy from bombardment by the molecules of the dispersion medium. It has been calculated that particles in Brownian Movement can change direction 10^{24} times per second. It is probably this movement associated with very small particles that maintains the stability of microemulsions.

This will be true, however, only if no coalescence occurs. In Chapter 5 a theoretical explanation is offered for thermodynamic stability that is based upon the existence of a transient negative interfacial tension associated with microemulsion systems.

E. Other

In addition to the foregoing measurements, there are a number of others which have utility for special applications. Included among these are conductivity, nuclear magnetic resonance (NMR), X-ray, electron microscopy, and light scattering depolarization. These are useful when the attributes of the microemulsion system measured by these means have significance in the final application. Sometimes, also, it is valuable to combine the results of these measurements on a single graph or table to obtain a comprehensive view of the system. A brief discussion of these measurements follows.

1. Conductivity

This is a simple measurement in which an electrician's ohmmeter or conductivity bridge is employed. It is important to provide a holder for the electrodes which keeps them at a fixed distance from each other in the emulsion.

Schulman measured the conductivity of microemulsion systems from the start but only in a qualitative way. He did not follow the change in conductivity as the systems inverted. Shah and Hamlin (12) did this, indicating, in addition, the changes in optical clarity and birefringence which occurred as the conductivity changed. This provided a method for determining at what point in the viscoelastic stage the aggregates change from cylinders of water to lamellar micelles.

Figure 1 illustrates these results. Molecular models of the
several forms of aggregates at the appropriate stages of
inversion have been superimposed on the Shah and Hamlin data
to obtain a simple overview of the course of the inversion.

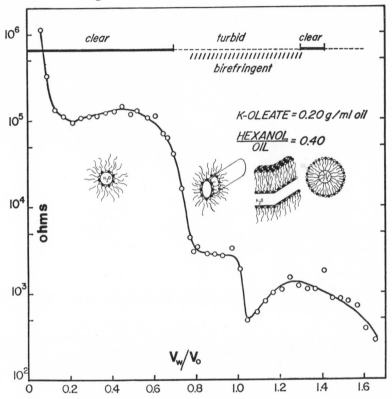

Fig. 1. *Electrical resistance vs. ratio of volume of water to oil in a microemulsion system during
inversion. Reproduced in part from "Structure of
Water in Microemulsions: Electrical Birefringence
and Nuclear Magnetic Resonance Studies," D. O. Shah
and R. M. Hamlin, Jr., Science, Vol. 171, pp. 483-5,
Fig. 1, 5 February, 1971. Copyright 1971 by the
American Association for the Advancement of Science.*

2. *Nuclear Magnetic Resonance (NMR)*

In general, NMR measurements are used to resolve theoretical questions regarding the state or location of molecules in
a microemulsion. For example, the line width of the NMR spec-
trum of the protons in molecules can be an indication of the
freedom of the molecules to thermal motion, the broadening of
the line indicating greater restriction of motion. Also the
chemical shift of water is different when it is distributed

in spheres or in cylindrical or lamellar micelles. In fact, Shah and Hamlin used high resolution NMR data to support their positioning of the cylindrical and lamellar micelles in the birefringent region. This was a practical as well as a theoretical contribution because it is now possible to locate cylindrical and lamellar micelles by conductivity measurements--a much simpler procedure. Birefringence and rheological measurements cannot do this as readily.

Other NMR studies have suggested that (a) the surface of the droplets in a water-in-benzene system is liquid but that the chains of the soap at the interface are somewhat constrained; (b) water molecules participate in a hydrogen bonded network at the water side of a soap stabilized o/w microemulsion in which the cation is 2-amino-2-methyl-1-propanol (AMP); (c) the packing conditions of the hydrocarbon chains of soap and alcohol determine the amount of water that can be held in a w/o microemulsion or inverted micelle (L phase); (d) the polar ends of the oleate soap are relatively immobilized at the aqueous interface in a water-in-benzene system while the terminal methyl end of the micelle is mobile in the benzene phase (cf. (a) above); (e) hexanol, in contrast, exhibits no motional restrictions, thus being free to partition between the interfacial film (interphase) and the benzene phase in order to equalize its chemical potential in each phase; and (f) a thin layer of water molecules of low mobility is associated with the polar heads of the surfactants at a w/o interface, but the core of water possesses the mobility of bulk water. These and other deductions from NMR studies are discussed and assessed in detail in Chapters 5 and 6.

3. X-ray

Low angle X-ray studies have provided some very pertinent information regarding the structure of microemulsions. Schulman used this tool to measure the diameter of both water droplets and oil droplets (13). He found that the water droplets gave stronger and less diffuse X-ray scattering than did the oil droplets. It is of particular significance at this time to note that with this technique he hoped "to bridge the gap between swollen micelles and emulsions" (14). Using this same tool, in a later paper (15) he provided evidence for systems composed of lamellar, cylindrical, and spherical micelles.

4. Electron Microscopy

A technique for staining organic materials containing double bonds with osmium tetroxide is available which enables one to take a picture of the dispersed phase of a microemulsion

directly, without shadow casting. Alkyd resins containing
linseed or soya oils, when dispersed into microemulsions,
serve as an excellent medium for the demonstration of this
technique. The micrographs in the frontispiece were made
in this way.

A dilute solution of an appropriate o/w alkyd microemulsion was exposed to the vapors of osmium tetroxide in a
vacuum desiccator. Almost immediately the droplets were converted into little cannon balls. When the dispersion was
placed on the stage of the electron microscope and vacuum
applied, the organic material burned off, leaving behind
osmium skeletons of the original spheres.

This technique is applicable only to unsaturated oils
such as alkyds or other materials of this kind in the form
of o/w microemulsions. When w/o emulsions of Carnauba wax
and carotene were treated in this way and sectioned, no pictures were obtained.

5. Light Scattering Depolarization

Light scattering depolarization is another way to detect
anisotropic aggregates in a colloidal system. In this case,
however, a Brice Phoenix or similar photometer is required.
This is a very accurate way to measure anisotropy and can be
used in micellar solutions to determine whether or not they
are anisotropic, even to a small degree.

IV. DEFINITIONS

With the foregoing as a background we are now in a position to approach a definition of a microemulsion and of a
micellar solution in a more disciplined way. Microemulsion
systems are considered to be two phase systems, whereas micellar solutions are considered to be one phase systems. This
is the basis of the present controversy. Even though this
difference in terminology appears to be only a question of
semantics, there may be a very real, although as yet un-
resolved, difference between these systems. The following
discussion presents some of the salient points.

Microemulsion systems contain oil, water, a surfactant,
and a cosurfactant (amphiphile). More explicitly, the stabilizing monolayer consists of the two components, a mixed film,
which film is probably penetrated to some extent by molecules
of the oil phase. Here the surfactants are considered to be
adsorbed between the two mutually insoluble liquids causing
one liquid to be spontaneously dispersed as microdroplets in
the other. This is the definition of an emulsion--a two
phase system.

On the other hand, many of the micellar solutions are
ternary systems although systems of four components have been
studied extensively. In this case, the surfactant or surfactants are viewed as spontaneously organizing themselves into
colloidal aggregates which can bind water or oil under the
appropriate conditions. Such systems are considered to be
one phase systems.

The difficulty is that the fluid isotropic regions,
whether w/o or o/w, in each of the systems exhibit the same
physical properties of light scattering, etc. It is fair to
say that the transition from the w/o to o/w in the microemulsion systems appears to follow a more consistent pattern
than in the micellar solutions.

This, then, is the crux of the dilemma. There is evidence to support both theories, and each will be presented by
its leading advocates. At this point in time there is no
overwhelming evidence to support one theory in favor of the
other.

When Schulman coined the term micro emulsion he used as
his frame of reference the emulsions with which he and Cockbain had worked (9). These were "fine" emulsions in the droplet size range of 0.5 to 4 microns and could be seen in the
light microscope. They scattered white light, i.e., they were
opaque like milk and separated on standing. On the other
hand, the dispersions he called micro emulsions did not separate and were transparent or translucent (opalescent). This
put the diameter of the particles below 1/4 λ, i.e., below
1400 A. Since these fluid w/o and o/w systems did not exhibit
optical, streaming birefringence, Schulman considered the
dispersed phase to be in the form of spherical droplets. He
measured the size of these droplets by the means available
to him at that time, low angle X-ray scattering, light scattering, and sedimentation velocity. In 1958, upon seeing
electron micrographs of spherical metallic skeletons of the
droplets of o/w alkyd emulsions, the diameters of which were
in the 75 to 1200 A range, he coined the term microemulsion
to describe these stable dispersions.

Such a description of these systems appeared adequate
for about a decade. During this period Shulman and his co-
workers developed the concept that the interactions in the
mixed film between the two mutually insoluble phases were
responsible for the properties of these systems. A critical
review of this work appears in Chapter 5.

In 1969 Friberg et al. (16) made the suggestion that
liquid crystalline phases could be in equilibrium with each
other and the isotropic solutions and that the so-called microemulsions could not be considered as true emulsions but rather
as solutions with solubilized water or solubilized hydrocarbons, i.e., one phase systems. Ekwall et al. (17) felt that

outside of the limit of the micellar range only two phase thermodynamically unstable emulsions could exist. His inferences were based on the study of ternary systems. Earlier, Adamson (18) had presented a model whereby w/o microemulsions were treated as systems of swollen aqueous micelles in which Laplace and osmotice pressures are balanced. This treatment, utilizing a four component system, was intermediate between the concepts of Schulman and those of Friberg and Ekwall.

The breach between the two concepts was subsequently widened by the Friberg school, which, utilizing excellently executed phase equilibria diagrams and painstaking analytical methods, seemed to indicate that one phase, fluid, isotropic systems could be obtained under much less rigorous conditions than for two phase microemulsions.

This is the way the matter stands at the present time. The outcome of this scientific debate will affect the formulator only in a minor way. For all practical purposes, either theory can and is being used to formulate stable, translucent systems of oil, water, and surfactants. The term "microemulsion" will be used to identify these systems, since both sides to the controversy use the term, although each ascribes a different meaning to it. The rest of the problem seems to be one of mechanics of formation.

REFERENCES

1. Becher, P., in "Emulsions, Theory and Practice" (P. Becher, Ed.), p. 2. Reinhold, New York, 1965.
2. Clayton, W., "The Theory of Emulsions and Their Technical Treatment," 4th ed., p. 1. The Blakiston Co., Philadelphia, 1943.
3. Selmi, F., *Nuovi Ann. d. Scienze Naturali di Bologna, Serie II, t.IV,* p. 146 (1845).
4. Moran, D. P. J., U. S. Patent 3,490,919.
5. Thomas, A. W., "Colloid Chemistry," 1st ed., p. 444. McGraw-Hill Book Co., New York and London, 1934.
6. Prince, L. M., in "Emulsions and Emulsion Technology" (K. J. Lissant, Ed.), Part 1, pp. 125-175. Marcel Dekker, New York, 1974.
7. LaMer, V. K., *J. Phys. Chem. 52,* 65, 68 (1948).
8. Wilton, I., and Friberg, S., *J.A.O.C.S. 48,* 771(1971)
9. Schulman, J.H., and Cockbain, E. G., *Trans. Faraday Soc. 36,* 551 (1940).
10. Schulman, J. H., Matalon, R., and Cohen, M., *Faraday Soc. Disc. 11,* 117 (1951).
11. Schulman, J. H., Stoeckenius, W., and Prince, L. M., *J. Phys. Chem. 63,* 1677 (1959).

12. Shah, D. O., and Hamlin, R. M., Jr., *Science 171*, 483 (1971).
13. Schulman, J. H., and Riley, D. P., *J. Colloid Sci. 3*, 383 (1948)
14. Schulman, J. H., McRoberts, T. S., and Riley, D. P. *Proc. Physiological Soc. 107*, 15 (1948).
15. Schulman, J. H., Matalon, R. and Cohen M., *Disc. Faraday Soc. 11*, 117 (1951).
16. Friberg, S., Mandell, L., and Larsson, M., *J. Colloid Interface Sci. 29*, 155 (1969).
17. Ekwall, P., Mandell, L., and Fontell, K., *J. Colloid Interface Sci. 33*, 215 (1970).
18. Adamson, A. W., *J. Colloid Interface Sci. 29*, 261 (1969).

Commercial History

LEON M. PRINCE

Consulting Surface Chemist
7 Plymouth Road
Westfield, New Jersey   07090

I.   INTRODUCTION

Since this is the first book to be written on microemulsions, there is an obligation to put the subject in historical perspective. It is quite fitting to do this because a chronological reconstruction of the work leads to an explanation of how the surface chemistry of these extraordinary systems evolved. As is often the case, the art preceded the theory. In our presentation in this chapter (and throughout the book), art and theory are blended with the express purpose of heightening our appreciation of the manner in which the industrial chemist and theoretician cooperated in solving practical problems in this field.

The first microemulsion was probably made by George Rodawald in St. Louis, Missouri (U.S.A.) in 1928. It was a dispersion of Carnauba wax-in-water. Professor Jack Schulman of Cambridge and Columbia universities did not identify such colloidal dispersions as unique until 1943, at which time he called the transparent water-in-oil (w/o) counterpart the

oleopathic hydro-micelle. These systems were not named microemulsions until 1958.

As one who has had experience with both the practical and
theoretical aspects of the subject, the author finds himself
in an unusual position to relate the early commercial history
and theory. He first became familiar with these systems in
1938 as a chemist with the Boyle Midway Division of American
Home Products Corporation. Doc Bowden and Jack Hohnstine of
that company had just developed a triethanolamine oleate Carnauba wax emulsion* which formed the basis for the Old English
and Aerowax self-polishing (no rub) floor wax formulas. Use
of this amine soap instead of Rodawald's alkali metal soaps
made the production of these emulsions by the inversion process much simpler. Earlier, the author had been in contact
with the manufacture of these Carnauba wax emulsions with
International Products Corp. of Trenton, New Jersey, in 1936,
where he had been unsuccessful in microemulsifying Candelilla
wax. Subsequently, he was involved in the microemulsification
of Ouricury wax, the synethic Carnauba wax replacements,
Chlordane, an alkyd and other materials. In 1958, he met Professor Jack H. Schulman, who held the chair of Stanley-
Thompson Professor of Chemical Metallury at Columbia University. Collaboration in the publication of two theoretical
papers on the formation and structure of microemulsions followed. The author has continued that interest.

It is of significance from a chronological viewpoint
that Rodawald's Carnauba wax emulsion preceded the commercial
nonionic emulsifiers like the Spans and Tweens by about ten or
fifteen years. These emulsifiers made microemulsification a
lot easier since their HLB could readily be varied over a wide
range that was compatible with the emulsifiability of the
oils. Soap, on the other hand, possesses an HLB value in the
range of 40, requiring low HLB materials like long chain alcohols to lower the HLB of their mixtures. Thus, soaps only
worked for a narrow range of materials that possessed some
alcohol or related amphiphile in their chemical makeup. To
this day, the chemical composition of oils remains a limiting
factor in the formulation of o/w microemulsions and to a lesser degree of the w/o types.

Carnauba wax was extraordinary in its ability to be
microemulsified. Rodawald's success undoubtedly was due in
large measure to his replacement of Beeswax with Carnauba in
his leather dressing formulas since Beeswax, by itself as the
oil phase, is not in Carnauba's class.

Since the art of making microemulsions of Carnauba wax is
not only important historically but the inversion process by
which they were originally made, may possibly be at the heart

*Cf. U.S. Patent 2,045,455 (1936). See also reference (1),
Chapter 6.

of the current controversy regarding definition, our narration will begin with a detailed description of how Carnauba wax emulsions came into being. The later developments will be put in their presumed chronological order.

II. CARNAUBA WAX EMULSIONS

As the story goes, Rodawald was interested in making a new and improved finish for the leather industry, a large segment of which was located in St. Louis. One day he made an emulsion of Carnauba wax and water that upon application to leather, dried to a glossy finish. It required no buffing to make it shine. This was so different from the emulsions that he had been making that he brought it home and tried it on his linoleum floor. It worked the same way there.

Being a practical man as well as an innovative chemist, Rodawald immediately recognized the possibilities of his discovery and went about exploiting them. Doing business as the Miracul Wax Company, he soon put on the market <u>Dri-Brite</u> floor polish. Not only did this start a new industry but a new kind of dispersion was introduced to colloidal chemistry. In spite of all the to-do about this product in the 1930's, there can be no refuting these facts.

Rodawald was making wax-in-water emulsions by melting wax in a pot, adding emulsifier to the liquid wax and then boiling water. This latter was done in many small aliquots. On the day he made the first microemulsion, the first water aliquot disappeared into the melt without any noticeable change in its appearance. So did the second and third aliquots. The mixture remained clear and fluid. At about the fourth aliquot, however, the mixture became caramel-like in consistency, although still clear. As more of the boiling water was added, the mixture in the pot became a gel and was so viscous that new water increments took a long time to become homogeneously blended into the mixture. But Rodawald persevered. By the time the volume of water exceeded twice the volume of the wax/emulsifier mixture, he was rewarded by a sudden thinning of the mixture in the pot. Upon dilution to about 15% solids, the consistency was as fluid as water and remained so upon cooling to room temperature. During the last stages of the water addition, the color of the emulsion changed from caramel color to a gray opalescent one, unlike a conventional emulsion, which is dead white.

The foregoing and what follows may be myth or truth; it is probably a little bit of both. It is hard to reconstruct laboratory events of half a century ago. Nevertheless, it puts Rodawald's discovery in perspective.

Normally, Rodawald would apply his finished emulsion to a

piece of scrap leather and buff it until it shone. Carnauba
wax imparted a higher gloss than did conventional Beeswax and
did so with less buffing. This accounted for his original
interest in Carnauba instead of Beeswax. However, as he continued to experiment with the Carnauba wax emulsions, he found
that by adding more soap (which he could do because soap was
cheap and the high melting point of Carnauba permitted it),
the emulsions became less milky and assumed a gray color.
This color change intrigued him. In fact, it was one of these
opalescent emulsions which conferred the gloss to his linoleum
without buffing.

This opalescence was the key to Rodawald's success with
his new product. As indicated in the previous chapter, opalescence is associated with droplet sizes smaller than 1400 A
or 0.14 μm. It was this small droplet size that enabled the
dried film to coalesce into a coherent or uniform film which
reflected light. And it was this same small droplet size
which made Rodawald's emulsion perform uniformly with time; it
was stable. The spontaneous development of gloss eliminated
the drudgery connected with the application of paste wax and
the stability feature soon attracted attention in other emulsion areas.

In order to make a product for linoleum or wooden floors
that worked better than his wax emulsion alone, Rodawald added
about 10% by volume of an ammoniacal dispersion of shellac.
This hardened the film and improved its gloss. Typical of the
Carnauba wax emulsion part of the formula used by Rodawald was
the following, in parts by weight:

Carnauba wax, prime No. 1	100
Oleic acid (red oil)	12
Potassium hydroxide	4
Borax (decahydrate)	7
Water	600
	723

An analysis of this product in terms of the surface chemical phenomena involved is of interest in the perspective of
the discussion in Chapter 5. Indeed, this formula and its
descendants helped in no small measure to mold the theory as
expressed there.

Rodawald was lucky (and this is not meant disparagingly)
in several ways. His substitution of Carnauba wax for Beeswax
was the first lucky step. Carnauba wax is one of the few
natural waxes which because of its chemical makeup (high hydroxyl value) is able to be microemulsified with soap. Beeswax can be microemulsified but with considerably more difficulty and usually not by itself. The choice of red oil was a
logical one; it was cheaper than stearic acid or coconut oil

fatty acids and it was liquid. However, had Rodawald used
sodium hydroxide instead of potassium, he might not have obtained his microemulsion. The potassium cation has a stronger
potential for making an o/w emulsion than has the sodium.
This is because, with its associated monolayer of water, this
cation has a larger area per moiety at the water side of the
interfacial monolayer, enabling this side of the film to expand more than the oil side. Next, he used borax. This was
a most fortuitous choice of ingredients. Borax decomposes into NaOH and boric acid at elevated temperatures. The boric
acid then liberates some soap to become free fatty acid, which
in turn acts as an alcohol in the interfacial monolayer. This
was very conducive to the formation of a microemulsion since
it essentially supplied more hydroxyl groups to the formula.

There was no luck in two other areas, and Rodawald must
be given full credit for his astuteness in these. His per-
severance in seeing the emulsion through the difficult-to-stir
viscoelastic gel stage cannot be underestimated in contribut-
ing to the success of his project. Finally, his keenness in
assessing the value of his new product was akin to genius.

The denouement of Dri-Brite was a sad one. Rodawald's
company was a small one, and he could not meet the competition
of the giants who soon copied his basic concepts. The result
was that the Dri-Brite formula was acquired by Boyle Midway in
1939. It was the author's sobering job to oversee its first
production. With the advent of World War II and the shortages
of raw materials, Dri-Brite dropped out of the market.

But Rodawald's brainchild was not without heirs. The
"emulsifiable" waxes of the 1940's and 1950's, which are discussed in the next chapter, were direct descendants of his
original Carnauba wax formula. Moreover, the concept upon
which Rodawald established his emulsion has found many imi-
tators in other fields.

III. CUTTING OILS

One of the areas where the concept of the stable Carnauba
wax emulsion found immediate utility was in cutting oils* as
coolants and lubricants for machine tool operations. Here the
improved performance of a stable emulsion could add immeasur-
ably to the performance of the product. As it turned out, the
makers of cutting oils added a few new wrinkles of their own
to the art of microemulsions.

It had been found that an emulsion of mineral oil and

---

*Rufus Rhodes, of the Sonneborn Division of Witco Chemical
Company, Inc., kindly assisted in the preparation of this
section.

water was very useful in high speed and alloy operations. The oil lubricated the cutting surface, the water imparted a much-needed coolant, and the emulsifier did double duty as an emulsion stabilizer and corrosion inhibitor. The emulsion was fed in a small stream to the point of contact of the tool and the work by means of a piping system and a pump. After sloshing over the working surface, the emulsion was recovered in a pan and recirculated. After a few such cycles, there was no longer uniform application of the ingredients, decreasing their efficiency. It was clear that a stable emulsion would be an improvement.

In the mid 1930's, it so happened that the Sonneborn Company was making Carnauba wax based floor polishes for the janitor's supply trade and was also a pioneer in cutting oil macroemulsions. Old art in one application became an innovation in another discipline. Development of a stable cutting oil emulsion did not take long once the idea came to mind. By the use of a coupling agent like diethylene glycol, an emulsifier in the soap class, and a petroleum sulfonate as a corrosion inhibitor as well as emulsifier, an elegant microemulsion system was developed which was stable in the recirculation system. By the late 1930's these systems were as commonplace as Carnauba wax emulsions.

These systems differed from Carnauba wax emulsions in several interesting ways. Because mineral oil is a liquid it was not necessary to heat the mixtures during formation of the microemulsion and the troublesome gel stage was avoided. Thus, at first, mineral oil was blended with petroleum sulfonate, soap, glycol and antifoam agent so that the user needed only to add water to obtain the microemulsion. Actually, because of the small amount of water in the petroleum sulfonate, this system was a clear w/o microemulsion. This was the "soluble oil" of commerce. Water could be added to it or the soluble oil added to water—a much easier process. This latter procedure required more emulsifying agent but since the emulsifier also acted as corrosion inhibitor, this presented no problem in this application. Later, and as practiced today, a soluble oil base is provided to the industry. This consists merely of petroleum sulfonate, soap, coupling agent, and antifoam. The user adds his own grade of mineral oil at concentrations to meet his needs. The resulting soluble oil is added to water to make the microemulsion.

These soluble cutting oils met government specifications and during World War II were used in large volumes.

IV. PINE OIL EMULSIONS

During the period in which the Carnauba wax and mineral

oil microemulsions were being introduced to the market, another microemulsion was in the process of being born. This was a pine oil emulsion and its development appears to have been made quite independently of the wax and cutting oil emulsions.

For many years pine oil had found use as a bactericide, fungicide, essential oil fixative and as a selective flotation agent. In the late 1920's, the Hercules Powder Company (now Hercules Incorporated) initiated a program to broaden the market for pine oil. One use that developed from the study was a "wetting agent" to enhance the action of detergents in commercial laundries for removing soil suspended in the water so that it could be rinsed out and not be redeposited on the clothes. This was a precursor of the modern anti-redeposition agents as incorporated in "built" detergent formulas. The product which Hercules introduced for this market in August of 1932 was called Daintex.*

This product found broad acceptance in the trade. Laundering formulas generally called for adding 10% soluble pine oil to the soap formula in the machine. In 1955, Hercules elected to withdraw Daintex from the market because its use as a laundry aid was by that time so well recognized that the company was beginning to compete with a large number of detergent formulators who were also purchasers of Hercules' pine oil.

The formulation of Daintex in parts by weight was:

Yarmor Pine oil	85.12
K or M Wood Rosin	4.21
Oleic Acid	4.21
Caustic Soda (50%)	2.21
Water	4.25

Daintex was a neutral, clear solution, making it a w/o microemulsion. It contained about 6% water and 9% soap, or about 10% soap based on the pine oil. Upon addition of this w/o microemulsion to the laundry machine, the emulsion inverted to a milky, opalescent o/w microemulsion.

It appears that the pine scrub soaps evolved from this system. In the late 1940's, the janitor's supply trade, which was closely allied to the suppliers of detergents for commercial laundries, picked up the idea of making an o/w emulsion of pine oil by simply diluting a formula like Daintex. By increasing the level of soap, it was possible to make a cheap, convenient product which could clean floors and walls as well as disinfect them. It was ideal for hospitals and large buildings. The odor of pine oil connoted cleanliness.

---

*The editor is indebted to Edwin C. Howard, Production Manager, Organics Dept., Hercules Incorporated, for his very considerable help with the preparation of this section.

These formulas as sold contained about 40% pine oil in concentrated form; they were diluted for use. Both rosin and fatty acids were used as emulsifiers for the pine oil. The concentrated product as well as the diluted one was clear and transparent. Obviously, the pine scrub soaps, as these products came to be called, were o/w microemulsions in the fullest sense of the term.

Formulation-wise, the pine scrub soaps had something in common with both Carnauba wax and cutting oil emulsions. Like the cutting oil emulsions, the gel stage was not a thick one that required considerable heat to manage. On the other hand, the pine scrub soaps resembled Carnauba wax emulsions in that they both utilized alcohol molecules from the oil phase as co-surfactants (cf. Chapter 5). Because of the much higher hydroxyl value of pine oil as compared to Carnauba wax it was not necessary to generate amphiphile *in situ* by the use of borax so that the pH of the pine scrub soaps could be in the 10.5 to 11 range. This was beneficial from the cleaning viewpoint.

When first introduced, the pine scrub soaps were considered to actually disinfect surfaces. This claim was ill advised in view of the large dilution of the pine oil. The fundamental concept underlying the success of these products was the fact that pine oil, which was water insoluble, was essentially made water soluble and very easy to apply. Its odor suggested disinfecting attributes which were more subjective than real.

With the tightening up of government regulation on bactericides and related agents, the janitor's supply trade has developed a Pine Disinfectant which is a clear o/w microemulsion of pine oil stabilized with soap. The Phenol Coefficients of these products are in the range of 3 to 5, depending upon grade. These could be called pine scrub soaps with high percentages of pine oil or a water-soluble (solubilized) pine oil.

The unusual flexibility of pine oil in entering into several microemulsion systems with ease reflects the propensity of the terpenes in general to microemulsify. The structure of these compounds appears to be very conducive to "emulsifiability" as defined in the next chapter.

V.   FLAVOR EMULSIONS*

During the 1940's microemulsions of flavor oils were

*Assistance with this section was generously furnished by Klaus Bauer of Dragoco, Inc. Consultations with Frank Fischetti of Fritzche-Dodge and Olcott are also gratefully acknowledged.

developed for the soft drink trade.   These were based on nonionic emulsifiers of the edible class.   Because of the high
ratio of emulsifier to oil, from 100% to 500%, such systems
could be and were called solubilized.   They were clear and
transparent.

A limited class of flavor oils were subjected to this
treatment.   The most famous one was oil of sassafrass, which
has long since been banned because of its adverse effect on
the liver.   Today usually only the citrus flavors are solubilized: orange oil, grapefruit oil, lime oil, lemon oil, etc.
These are used in the beverage and cola industry as well as
being incorporated in foods.

Flavors for mouthwash are still microemulsified, as is
lemon oil, utilizing the usual ethylene oxide adducts.   Interest is not lacking in these areas as witness a recent article
in the March, 1975, *Dragoco Report* entitled "The Theory and
Technology of Solubilized Systems" from their Italian labo-
ratories.

VI.   PESTICIDE EMULSIONS

After the development of DDT, there appeared on the market a number of insecticides or pesticides based on chlori-
nated moieties.   One of these was Chlordane, octachlor indene,
which came on the market in the late 1940's.   It was a viscous
liquid.   This product was useful in the agricultural field but
was also very useful in killing termites and cockroaches.   It
was in these latter areas that water-based microemulsions of
Chlordane found utility.

Stability during storage and application were important
pluses to pest control operators and exterminators so that the
premium price for microemulsions was justified.   The water-
based emulsions also avoided the use of flammable, toxic and
smelly solvents.

The first Chlordane microemulsion was made by blending
Chlordane and Atlas Powder's G-8916P nonionic emulsifier in
equal parts.   When this blend was poured into water, a clear,
transparent dispersion resulted.   It was called a solubilized
system since the term microemulsion had not yet been coined.

Later, the writer made opalescent o/w emulsions consisting of much less nonionic emulsifier.   These did not wash
away so quickly and were cheaper.   A soluble oil was also made
of emulsifier and chlordane which needed only to be added to
water to make a stable o/w emulsion.   These products were mar-
keted under the trade name of Lucide.

The original emulsion made by the author was a copy of
the soap-stabilized Carnauba wax emulsion.   The chlordane
formed the same kind of opalescent microemulsion but was not

chemically stable. The alkalinity of the soap soon dehydro-
genated the chlordane and the HCl so formed, gelled the soap.
As a result, the first demonstration was a disaster. It was
obviously necessary to use nonionics.

The similarity in the performance in the microemulsions
of Carnauba wax, Pine oil and chlordane, ostensibly quite different chemically, impressed the author. Later in his collaboration with Professor Jack Schulman, this equivalency
formed the basis of some fundamental theoretical concepts.

VII. EMULSION POLYMERS

Small droplet size dispersions of polymers formed by the
emulsion polymerization process are microemulsions in the same
sense that Carnauba wax emulsions are. Both begin with a liquid "oil," water and emulsifier and end up with solid spheres
dispersed in water. It is true that emulsion polymerization
mechanics differ from conventional emulsification processes
but, as we shall see, by not too large a margin.

Immediately after World War II in the late 1940's, the
styrene-butadiene latices used for making synthetic rubber
were adapted to the paint industry. A water-based vehicle was
developed which was an aqueous dispersion or latex of a
styrene-butadiene copolymer the droplet size of which was
greater than $\frac{1}{4}\lambda$. These water based paints were an instant
success. Combined with pigment, color and other adjuvants,
this paint was easy to apply and very easy to clean up after.
Brushes could be washed with water and soap, and spills could
be washed away with water.

As improvements were made in this new art, it was found
that latices could be made, the droplet size of which were
less than $\frac{1}{4}\lambda$ so that films were more coherent. With a small
amount of plasticizer, films of such latices could be laid
down on vinyl tile and wooden floors and reflected more light
than the wax/shellac polishes. They were tougher also, giving
longer wear. Most importantly, such products made the American manufacturers essentially independent of foreign supplies
of wax. Although some wax emulsion was used in these products
for a number of years, synthetic emulsifiable waxes became
available also, so that Carnauba wax came to have application
only in certain specialized uses. Today microemulsion polymers completely dominate the floor polish industry.

Only a number of monomers and combinations of them yield
the small droplet size latices suitable for the floor polish
industry. These are chiefly the acrylics and styrene. Usually only a few percent of emulsifier is required to emulsion
polymerize these monomers into latices having droplet sizes
less than $\frac{1}{4}\lambda$. This is an improvement over the 20% needed to
microemulsify wax.

That there is some connection between the microemulsion process and emulsion polymerization process is demonstrated by the following experiment. This was discussed in Chapter 1 from the viewpoint of special light scattering effects. Vinyl acetate is conventionally emulsion polymerized with a few percent emulsifier to yield latices for paint and adhesives. The latices are dead white, i.e., their droplet sizes are in the micron range. By first making a clear and transparent microemulsion of vinyl acetate monomer in water using 40% Pluronic F-68 on the weight of the monomer and then emulsion polymerizing this microemulsion with catalyst, an extremely uniform droplet size opalescent latex was made which scattered light of several colors. This was certainly not an economical process but demonstrated a connection between microemulsification and emulsion polymerization. It would appear that in the latices, polymer molecules are adsorbed to and are oriented in the interfacial monolayer and occupy a considerable volume of it (cf. Chapter 5).

VIII. OTHER SYSTEMS

Since 1955 microemulsions have appeared in a wide range of fields, frequently without the knowledge of their formulators that their products were microemulsions. A number of these are mentioned here.

Probably the most widely used microemulsions outside of latices for paint and floor polish are the w/o microemulsions used in dry cleaning establishments. In these, a small amount of water is incorporated in the dry cleaning fluid to impart both water and oil solubility to the cleaning product. These systems have been in use since the late 1940's and are clear and transparent.

Another w/o system that has great potential for high volume use is the microemulsion slug being proposed for tertiary oil recovery. At its maximum potential, this slug is capable of recovering as much oil from the ground as has already been taken out! The subject is discussed in some detail in Chapter 7.

A product that involved several of the principles involved in the formation of microemulsions was an alkyd emulsion formulated in 1955 for Reichhold Chemicals, Inc. It was called Synthemul 1505 and offered a solvent-less vehicle to the trade. It is still being sold. It was this emulsion and modifications of it from which the micrograms of the frontispiece were made.

To date a number of physiological applications of microemulsions have surfaced. Two patents, one U.S. No. 3,911,138 and a second, German No. 2,319,971, have been issued for perfluorocarbon emulsions to serve as intravascular oxygen and carbon dioxide transport agents. The function of the micro-emulsion is for stability purposes but the whole scheme is still in the experimental stages (1,2). Such systems may have a big future as artificial blood. In another area, microemulsion systems including the liquid crystalline phase between the o/w and w/o types have been utilized to study the clinical potency of anesthetics (3). Microemulsification has also been suggested as an alternative to micellization of fat in the intestines by bile acids (4). It has been proposed that micellar solubilization can make drugs which are soluble in oil easier to administer by making water the external phase (5).

A means of producing microemulsions by an ionic pumping action has been disclosed in U.S. patent No. 3,813,345. The process is a long and complicated one, illustrating the trouble that one will resort to in order to obtain a microemulsion.

As previously indicated, nonionics of the polyoxyethylene class serve as ready agents for microemulsification and many uses for them have been found (6). Some of these are listed: solubilization of vitamins and essential oils; emulsifiable solvent cleaners for metal and paint brushes; ringing gel hair pomades; solubilized perfume for hair dressings; a baby shampoo; alcohol-less clear colognes; waterless hand cleaners; after-shave lotions; bath oils; emollient body preparations; hair styling agents; and mineral oil lotions.

REFERENCES

1. Clark, L. C., Jr., Becattini, F., and Kaplan, S., *Triangle 11*, 115 (1972).
2. Rosano, H. L., personal communication.
3. Shah, D. O., *Ann. New York Acad. Sci. 204*, 125 (1973).
4. Prince, L. M., "Biological Horizons in Surface Science" (L. M. Prince and D. F. Sear, Eds.), pp. 353-366, Academic Press, New York, 1973.
5. Dittert, L. W., *in* Sprowl's "American Pharmacy," pp. 150-152, Lippincott, Philadelphia, 1974.
6. Becher, P., private correspondence.

Formulation

LEON M. PRINCE

Consulting Surface Chemist
7 Plymouth Road
Westfield, New Jersey   07090

I. INTRODUCTION

In this chapter every effort is made to provide the bench chemist and theoretician with the tools needed to solve everyday problems in microemulsions and to design new products. The techniques available to the formulator will be described. In addition to the Hydrophile-Lipophile Balance (HLB), a number of alternate and complementary techniques that have been developed in recent years and have been found useful, will be presented. Some of these are theoretical in nature and some practical. These techniques will be discussed in detail and illustrated by examples. Art and theory will be integrated at the formulation level.

The formulation of microemulsions or micellar solutions, like that of conventional macroemulsions, is still an art and practiced as such. In spite of reasonably precise theories which explain the Physics and Chemistry of their formation and behavior, the science of microemulsions has not advanced to the point where one can predict with accuracy what is going to happen in the beaker, in the reaction vessel, or in the ground (tertiary oil recovery), with all mixtures of ingredients.

The very much higher ratio of emulsifier to disperse phase that differentiates the microemulsion from the macroemulsion, deceptively appears to make the application of the various techniques less critical, at least in the early stages of the development of a formula. But by the time the final stages are reached the requirements of the microemulsion emerge as much more critical because of the greater number of parameters which must simultaneously be met.

Microemulsification is concerned with the stable dispersion of oils in water and water in oils. To date, no microemulsions have appeared in which one of the mutually insoluble liquids is not water. Our attention will therefore be directed to matching emulsifiers to oils in such a way as to produce o/w and w/o emulsions of small droplet sizes, i.e., less than ¼ λ. The role of water in the matching process is still incompletely understood. What is known about it is discussed in Chapter 5, and its practical implications are utilized in the formulation context here.

At the outset, it must be explicitly stated that there are limitations to the nature of the oils that have been microemulsified, and there are always product specifications which decrease the range of emulsifying agents that can be employed in given formulas. Few oils in their natural form seem to be chemically constituted to form these stable systems with water. This is true of both o/w and w/o types but is particularly so of the o/w types. Up to this time, the author knows of only about 50 o/w commercial microemulsions. Most oils do not microemulsify regardless of how much excess emulsifier is employed. Means to minimize these difficulties are discussed under "Emulsifiable Oils."

It is true that in recent years the number of these transparent or translucent systems, particularly of the w/o types, has been increased by the investigators of micellar solutions. These systems appear on phase equilibrium diagrams, but it is fair to say that it is generally not feasible nor desirable to reduce them to practice. Many of these systems will be dealt with by Professors Shinoda and Friberg in the later chapters.

As the first chapter was devoted to the Physics of these colloid systems, so this chapter will be devoted to the Chemistry of the molecules which make up the systems. Actually, it will be concerned primarily with the physical interactions among these chemical species and only secondarily with their chemical interactions. Emulsions are a branch of Colloid Science called surface chemistry, and the subdivision of surface chemistry is, specifically, interactions at the liquid/liquid interface.

II. MECHANICS

The mechanics of forming microemulsions differ somewhat from those used in making macroemulsions. The most significant difference lies in the fact that putting work into a macroemulsion or increasing emulsifier content usually improves its stability; not so for microemulsions. These systems appear to be dependent for their formation upon specific and as yet incompletely known interactions among the molecules of oil, emulsifiers, and water. If the specific interactions are not realizable, no amount of work input nor excess emulsifier will produce the desired product. Ultrasonication, high speed or high shear homogenization is to no avail if the chemistry is not right. On the other hand, when the chemistry is right, microemulsification occurs (almost) spontaneously. There are several ways of blending the ingredients of microemulsions. Before discussing these it should be remembered that these systems consist of at least 10% emulsifier on the weight of the oil; usually, 20%-30% emulsifier on the weight of the oil is present. Moreover, the techniques for w/o systems are simpler than those for the o/w systems in keeping with the greater difficulty of finding the proper match between oil and emulsifier for the latter systems.

The w/o systems are made by blending the oil and emulsifier, with a little heat if necessary, and then adding water. The amount of water that can be added to a given system of emulsifier and oil may not always be high enough for the application in mind. In that event, it becomes necessary to try other emulsifiers. When one is found that permits of the desired water uptake, it may be convenient from a processing

viewpoint to add the mixture of emulsifier and oil to the
water. Again, warming the system may speed the mixing process.
It is axiomatic that in systems of oil, water and emulsifier
that are capable of forming microemulsions, the order of mixing does not affect the end result. This is also true of the
micellar solutions described by Professors Shinoda and Friberg
in the later chapters.

The order of mixing for the o/w systems is open to a
wider range of options. Some may be better for one system
than another, but in no case can a microemulsion be formed
unless the proper match between oil and emulsifier exists.
Perhaps the simplest way to make an o/w microemulsion is to
blend the oil and emulsifier and then pour this liquid mixture
into the water with mild stirring. In the case of waxes, both
the oil/emulsifier blend and the water must be at high temperatures. Indeed, with waxes whose melting points are above the
boiling point of water, the mixing must be done in a pressure
vessel to prevent the wax from freezing during the emulsification process. Another technique is to make a crude macroemulsion of the oil and one of the emulsifiers, for example, a
soap. By using low volumes of water a gel is formed. This
gel is then changed into a clear sol by titration with a second surface active agent like an alcohol. This system may
then be transformed into an opalescent o/w microemulsion of
the desired concentration by further addition of water (1).
By far the most common method of making an o/w microemulsion,
especially in the trial and error stage, however, is by the
so-called inversion process. This is described below in some
detail because of its importance.

In actual practice, oils which are capable of being microemulsified, i.e., "emulsifiable oils," as opposed to those
which may be dispersed in micellar solution, invert by the
slow addition of water from a fluid w/o dispersion through a
viscoelastic gel stage to a fluid o/w microemulsion. This
pattern is so well defined as to be almost a hallmark of the
o/w microemulsion. As discussed in Chapter 1, the viscoelastic gel stage consists of cylinders of water adjacent to
the w/o dispersion and lamellar micelles adjacent to the o/w
microemulsion. Their optical properties in combination with
their rheological behavior make identification positive without instrumentation.

Accordingly, this method of preparation is the preferred
one for initial exploration. Usually, 100% emulsifier on the
weight of the oil is employed. After carefully blending--
with heat if necessary--water is added to the blend in a beaker. This is done in successive, small aliquots. If the
chemistry is right, a clear, transparent w/o dispersion first
forms. This is fluid. As more water is added, at about equal
volumes of water and oil/emulsifier blend, the system begins

to become more viscous.   As more water is added, it becomes
very viscous, ultimately becoming a heavy gel.   At this point
it is frequently helpful to apply heat to thin the gel and
facilitate passage through this stage.   With the addition of
more water, the gel eventually thins out to a fluid o/w microemulsion which can readily be identified by its clarity or
opalescence.

The highly viscous intermediate gel stages are obviously
not microemulsions but are sometimes so called, as in the case
of ringing gels used as hair pomades, etc.   These systems are
actually liquid crystalline phases and occur because of the
particular sequence of mixing employed in forming the microemulsion.

The appearance of the highly viscous stage which may be
clear or opalescent is good evidence that the formulator is
close to matching his oil and emulsifier.   Unfortunately, in
many systems a clear w/o dispersion forms at first and begins
to pass into the gel stage but fails to invert to a fluid o/w
microemulsion.   This means that the match between the oil and
emulsifier is not quite good enough.   Adjustments in keeping
with the suggestions made in subsequent sections are then in
order.   In other cases, the system may pass through the viscoelastic gel stage and form an o/w microemulsion that is too
viscous for practical use or its particle size is too large as
noted from the light it scatters.   If the emulsion is too viscous, the HLB of the emulsifier system should be slightly increased; if the emulsion particle size is too large (and the
system potentially unstable), the HLB of the emulsifier system should be lowered.   The change in HLB can be effected in
a number of ways consistent with the requirements of the finished product.

According to the above procedure, it should not take too
many trials with 100% emulsifier on the weight of the oil to
find a system with a small volume of water (a fraction of the
volume of the oil) that is clear or translucent.   It is when
the percentage of emulsifier is decreased that the real test
comes.   On the other hand, if the system can be carried
through the gel stage, a microemulsion can usually be brought
home.   Once this is done, the inversion process may be aban-
doned since it is cumbersome, although it utilizes less emulsifier than other methods.

III.   CHOICE OF EMULSIFIERS

It would appear of the utmost importance for the formulator of microemulsions to have no preconceived notions concerning the class of emulsifier he is looking for.   The knowledge
of molecular interactions which can take place in two

dimensions is meager at best. Because of this it often happens that unexpectedly good results are obtained by intuition or luck with surfactants which, on the face of things, would not be expected to interact beneficially for microemulsification purposes. Thus, it is well to take a good look at the list of emulsifiers in McCutcheon's "Detergents and Emulsifiers" (The Allured Publishing Corporation, Ridgewood, New Jersey 07450). The listings in this annual usually contain the molecular structure or, at least, the type of the emulsifier and its HLB. It is not a complete list. For example, soaps are not included, all manufacturers are not included, and the manufacturers that are included do not necessarily list their complete lines. Nevertheless, this listing is complete enough to suggest many possibilities to the formulator, and it is in this broadest of perspectives that the formulator should begin his task.

Of course, there are definite limitations to this admonition. A food product requires edible emulsifiers, and a halogenated hydrocarbon should not be emulsified with soap lest dehydrohalogenation occur, etc. Other prerequisites of the final product such as odor, color, taste, or price will impose additional limitations. In spite of these, the list is long enough and the combinations of two or more emulsifiers are almost endless.

There are a number of emulsifier selection systems (2). In this discussion four will be considered: the Hydrophile-Lipophile Balance (HLB), the Phase Inversion Temperature (PIT), the Cohesion Energy Ratio (CER), and the partitioning of cosurfactant between the oil phase and interphase, $(\gamma_{o/w})_a$. Each of these systems stands by itself, but they are interrelated by means of the original HLB concept. Their employment in combination materially assists in matching the chemical type of the emulsifier with that of the oil.

The use of the phase equilibria diagram as a tool to match oil and emulsifier or to minimize the amount of emulsifier needed to effect microemulsification is not discussed in this chapter. Professors Shinoda and Friberg consider these aspects of the subject in their chapters.

A. The Hydrophile-Lipophile (HLB) System

Given an oil to be microemulsified, the formulator should first determine its required HLB. This is done in the same way as for macroemulsions (2,3,4), giving two values. The lower one in the range of 4-7 is for w/o emulsions, and the higher one in the range 9-20 is for o/w emulsions. This is usually done experimentally, but, if one is lucky, the required HLB may be found in Table 2 of Reference (3). With this HLB number, one then must try to find the chemical type

of emulsifier which best matches that of the oil. This is
what makes the microemulsion. Naturally, hydrophobic portions
of surfactants which are similar to the chemical structure of
the oil should be looked at first. Since polar groups on the
emulsifier also play an important role in the chemistry of
the surfactant, they must also be considered in the matching
process. Sometimes, emulsifiers of the needed HLB are not
regularly supplied by the manufacturer. In this case arrange-
ments can usually be made to accommodate the formulator.

A short review of the origins of the HLB emulsifier selection scheme may be helpful in defining the limitations of
its capabilities. It began with the recognition that the
properties of the sorbitan esters which were made water soluble by adding ethylene oxide depended upon their apparent solubility in water. This solubility behavior was found to be
dependent upon the weight ratio rather than the mol ratio of
ethylene oxide content, in this way compensating for the degree of hydrophobicity of the tails. Then it was realized
that not only solubility but emulsion behavior was dependent
on this weight ratio. Since it was recognized early in the
game that ionic surfactants possessed HLB's much higher than
even the theoretical maximum for nonionics, the weight percentage of hydrophile was arbitrarily divided by 5 and the resulting number used as the classifying system.

Such a classification system completely disregarded the
interaction of the emulsifier with the oil; only water solubility was taken into consideration. This left matching of
the chemical type of the emulsifier to that of the oil open to
trial and error, albeit on a much less extensive basis. It is
this deficiency in the HLB scheme that the other selection
systems attempt to make up for.

Since the HLB scheme was primarily designed for nonionics
of the ethylene oxide class, these will be discussed first.
In the total scheme of emulsifiers, these compounds have certain advantages and disadvantages. First of all, they are
relatively inert to hard water as compared to soap, they are
inert chemically in most situations, and they can be made edible. At the time they were introduced, these were fine sell-
ing points. Along with a reasonable price, abetted by the low
cost of ethylene oxide, it quickly put these materials on the
market to stay. Among their disadvantages were the fact that
these materials had a wide (Poisson) distribution of polyoxyethylene homologues. This was superimposed on the natural
wide distribution of fatty acid species in any given hydrophobic moiety. In some cases, this heterogeneity can be made
use of in improving emulsifiability, although Shinoda feels
that a homogeneous emulsifier is the better actor (cf. next
chapter). The most serious shortcoming of the ethylene adduct
nonionic emulsifier is its negative temperature solubility.

Changes in temperature have a large effect on the efficiency (and stability of systems made) of these emulsifiers so that great care must be exercised in their use. Probably the most pragmatic way to overcome this shortcoming is to employ more than one kind of nonionic so that their emulsifying efficiency extends over a wider range. But this requires an extensive testing program.

Although the ethylene oxide adduct nonionics are the most important class, there are other nonionics which find a place in microemulsion formulation. These are the polypropylene or butylene oxide adducts and combinations of these with each other and ethylene oxide. Derivatives containing polar groups of sulphur, nitrogen, and phosphorus are also popular. Finally, there are the alcohols such as the long chain normal aliphatic or methyl cyclohexanol which act as cosurfactants to reduce the high HLB of soap or detergents like sodium cetyle sulphate. These cosurfactants have very low HLB's, and with other cosurfactants like cholesterol or long chain amines, which are used with cationics, serve to lower interfacial tension between the two mutually insoluble liquids. This is a very effective device as first demonstrated by Schulman and Cockbain (5) in their famous paper. This is extensively discussed on a theoretical basis in connection with microemulsions in Chapter 5 and in a quasi-scientific way in the appendix of this chapter.

Suffice it to say here that in the formulation of a microemulsion, two surfactants are almost always employed. One is called the surfactant and is usually water soluble or dispersible such as soap or a detergent; the other is called the cosurfactant and is an alcohol or low HLB nonionic. The effect of combinations of surfactant and cosurfactant on their emulsifying potential can be very subtle as seen below.

The formation of cosurfactant *in situ* is a case in point. As previously indicated anionics (and cationics) have high HLB's. Griffin assigned a value of 40 to sodium lauryl sulfate. Soaps have HLB's less than 40 but more than 20 depending upon the composition of their fatty acid portions as well as their cations. Normally, to bring a soap into the emulsifying range, one would use a long chain alcohol or low HLB nonionic in combination with it. However, the HLB of soap is pH dependent. At pH 10.5 or higher, all the fatty acid exists as carboxylate ion; at pH 8.8, half the fatty acid is free, acting as an alcohol at an emulsion interface; and at pH 6.8, 2 mols of the fatty acid are free and one is still a carboxylate. The effect of "acid" soap on its HLB is illustrated by the following example.

Formula A: Kerosene or Dodecane   20 g
               Oleic acid (red oil)   4
               2-Amino-2-methyl-1-propanol (AMP)   2.25
               Water (deionized)   160

The soap is dissolved in the oil and the water is added in small aliquots while maintaining the temperature at 70°C. The first addition of water turns the clear solution of soap and oil into a milky white w/o dispersion. When the volume of water added is equal to the volume of the oil, the system becomes slightly viscous and as more water is added, it inverts to an o/w macroemulsion that is milky white.

If, however, 2 grams of cetyl alcohol is added to the soap and oil or 0.5 gram of boric acid is added to the first addition of water, a clear w/o dispersion is first formed which upon further addition of hot water passes through the viscoelastic gel and finally inverts to an o/w microemulsion. The cetyl alcohol lowers the HLB of the soap to that of the required HLB of the oil phase and the boric acid drops the pH of the soap solution to 8.8 so that one-half of the soap is in free fatty acid form which, in this instance, acts like oleyl alcohol, lowering the HLB of the combined emulsifier system so that a microemulsion can form.

Another subtle example of how emulsifiers combine to produce required HLB's for oils is in ethylene oxide adduct nonionic systems. As indicated, there is a wide distribution of homologues in a commercial nonionic. Thus, one will find very low and very high HLB fractions as well as a major fraction of the compound in the average HLB range for a given nonionic. The low HLB fraction finds a very special use in microemulsification. It acts like an alcohol and distributes itself between the oil phase and interface (or interphase) so as to substantially lower the original interfacial tension between the oil and water. This will be discussed more fully in the section dealing with $(\gamma_{o/w})_a$.

Because nonionics, in general, are less efficient, pound for pound, than anionics or cationics, they are often mixed with ionic emulsifiers. In a way, this solves the surfactant, cosurfactant problem in a single stroke. The anionic is the high HLB partner and the nonionic, the low HLB one. For microemulsification, considerable specificity exists so that only certain combinations of nonionics and ionics are effective. Experience and trial and error are the only answers to what combinations are right for your particular oil.

In the case of anionics, a great deal of work has been done. Although the HLB of anionics is not generally known as it is for nonionics, their lower cost has spurred a great deal of effort on them. At the very outset, Schulman and McRoberts (6) found that with soap as the surfactant, the continuous phase was water when the alcohol cosurfactant was shorter than seven carbon atoms, and oil was the continuous phase at higher molecular weight alcohols. This was true of mineral oil and water emulsions. The more sophisticated microemulsion cutting oils (mineral oil-in-water) utilized petroleum sulfonates, rosin and/or fatty acid soaps and a coupling agent in the form of ethylene, propylene, or hexylene glycol. The rationale behind this formulating device is that the water soluble coupling agents behave as oil soluble cosurfactants. In this way they effectively lower the HLB of the soaps to that of the mineral oil, the required HLB of which is in the range of 10 to 12, depending upon whether the oil is paraffin or aromatic. Petroleum sulfonates also have a lower HLB than the usual alkyl sulfate, such as sodium lauryl sulfate, simplifying the matching process.

As a final word, the reader should be reminded that as with petroleum, so vegetable and animal oils and fats are on the verge of being denied to us. In 1975, the world population was estimated at four billion souls; in the year 2000, it may be seven billion. Moreover, American cattle growers are experimenting with ways to grow beef animals that have much less fat (tallow) than the current models. This points to a situation where oils and fats will be forbidden for use other than in food. Surfactants from other sources, therefore, should be considered even in the context of today's formulations. Among the hydrophobic moieties that look promising for the long haul are derivatives of lignin, cellulose, sugar, rosin, petroleum-derived alcohols, and fatty acids (not 100% linear), alpha olefins, alkyl benzenes, lecithin, lanolin, naphthalene, petroleum sulfonates, block polymers of ethylene and propylene oxide, polymers in general, and even inorganic derivatives as from clay. Moreover, any and all of these must be considered in the context of environmental safety.

B. The Phase Inversion Temperature (PIT) System

Shinoda (7) proposed an emulsifier selection system based upon the temperature at which an emulsifier causes an o/w emulsion to invert to a w/o emulsion. He has called this the PIT system and, unlike the HLB system, it provides information concerning the types of oils, phase volume relationships, and concentration of emulsifier. It is established on the proposition that the HLB of a nonionic surfactant changes with temperature and that the inversion of emulsion type occurs when the hydrophile and lipophile tendencies of the emulsifier just balance. No emulsion forms at this temperature. Emulsions stabilized with nonionic agents are o/w types at low temperatures and invert to w/o types at elevated temperatures. Shinoda has also designated his PIT system the HLB-temperature system.

From a microemulsion viewpoint PIT has an outstanding feature. It can throw light on the chemical type of emulsifier needed to match a given oil. Thus, the scheme provides an experimental basis for chemical matching when intuition

and all else fails. Indeed, the required HLB values for various oils estimated from the PIT systems compare very favorably
with those prepared by the HLB system as presented in Table 2
of reference (3). Professor Shinoda has presented an extensive description of the operational potentials of the PIT system in the next chapter.

C. The Cohesive Energy Ratio (CER) System

Recently, a fundamental basis for the HLB concept has
been developed by Beerbower and Hill (8,9) which is applicable
to nonionic as well as many anionic emulsion systems. It has
been called the "Cohesive Energy Ratio" or CER system for
short. By means of an equation based on thermodynamic parameters an emulsifier can be found to match a given oil to a
degree not possible by the cruder HLB scheme.

The HLB concept as commonly used is based on the weight
fraction of hydrophilic material in the surfactant, without
allowing for degrees of hydrophilic strength in the heads and
lipophilic activity in the tails. This leaves to the formulator the task of "chemical matching" the tail of the emulsifier to the oil by trial and error. The resultant difficulties are well illustrated in Figs. 1 and 2 of reference (8).

CER combines the theoretical but nonnumerical Winsor concept of R (the ratio of dispersing tendencies) with the London
cohesive energies developed by Scatchard and Hildebrand (Solubility Parameters) at the oil side of the interface with the
London, Keesom (dipole), and hydrogen bonding cohesive energies developed by Hansen at the water side. The result is an
extrapolation to macroemulsions of Winsor's R which was originally devised to explain the phase relationships in "solubilized" systems. If these solubilized systems are considered
microemulsions, then CER is uniquely qualified to be used as
a matchmaker for oil and emulsifier.

The master CER equation, Eq. (9) of reference (8), applies to all emulsion formulations regardless of the value of
the HLB or whether or not the tail of the emulsifier matches
the structure of the oil, making the use of the equation difficult at best. However, it can be simplified into a much
more useful form by excluding all poorly balanced emulsions.
This leaves a concise mathematical expression of any well-
formulated emulsion in that CER equals the ratio of head volume to tail volume multiplied by the square of the ratio of
the solubility parameters. Since the emulsifier now effectively blends water and oil, the properties of these phases
need not explicitly be excluded. Thus, the ratio of volumes
is simply another expression of HLB while the ratio of parameters constitutes the chemical match. If the solubility
parameter, molecular weight, and density of the oil are known,

Eq. (11) of reference (8) is easily used to calculate its HLB requirement for either o/w or w/o formulations. Alternatively, the surface tension, molecular weight, and density may be used for this purpose, in which case Eq. (13) of reference (8) is applicable.

Beerbower has indicated that the CER system as presently constituted is only designed to handle emulsions as small as 5 μm in diameter. Below this size, droplet curvature affects the applicability of his equations. In view of Beerbower's ability to match oil to emulsifier tail, his system commands attention because it provides a means, other than the trial and error method, to find matches which produce microemulsions.*

D.   Cosurfactant Partitioning $(\gamma_{o/w})_a$

In order to explain the spontaneous formation of microemulsions, the thermodynamic equation associated with film balance studies on the Langmuir trough was resorted to. This is concerned with the molecular interactions in the monomolecular film which envelop the dispersed phase of the emulsion. A description of how this monomolecular film is formed and the basis for its effect on microemulsification is contained in an appendix to this chapter. As far as possible this dissertation is couched in terms that are meaningful to the bench chemist. The more complete account is presented in Chapter 5.

For our purposes here, it is sufficient to relate that much consideration had been given to the influence of the chemical nature of both the alcohol and oil phase upon the formation of microemulsions stabilized with soap and long chain aliphatic alcohols. It was inferred that the condition responsible for the spontaneous formation and stability of these small droplet sized dispersions was a zero interfacial tension. This could come about according to the equation

$$\gamma_i = \gamma_{o/w} - \Pi$$

in which γ_i is the total interfacial tension, $\gamma_{o/w}$ is the interfacial tension before addition of stabilizing agents, and Π is the two-dimensional spreading pressure in the monolayer of adsorbed species, all providing $\Pi = \gamma_{o/w}$. Since the o/w interfacial tension between a n-paraffin and water is about 50 dynes/cm, this would mean that the spreading pressure among the film tenants would have to be of the same magnitude. As it turned out values of Π of this magnitude had been measured on the Langmuir trough.

---

*The editor gratefully acknowledges the cooperation of Alan Beerbower in the preparation of this section.

What was more to the point was that values of Π <u>in excess of</u> 50 dynes/cm had been measured. This led to the suggestion that a transient, <u>negative</u> interfacial tension was responsible for microemulsification. Equilibrium would be attained when the transient tension returned to zero by virtue of the uncrowding of the interface tenants and loss of pressure in the interface. The prime mover in the development of the high values of film pressure was considered to be the penetration, on a highly selective basis, of molecules derived from the oil phase into the interphase.

Modification of this theory was required when it was determined that at two-dimensional pressures in the range of 50 dynes/cm, hydrocarbon molecules would be ejected from the monolayer. It was then proposed that the initial negative interfacial tension, γ_i, was the result not so much of a high value of Π but of a large depression of the value of $\gamma_{o/w}$, the original interfacial tension between the oil and water before the emulsifiers were added.

Since most microemulsions only appear to form readily in the presence of a cosurfactant which is oil soluble, it was submitted that this material distributed itself between the oil phase and interphase and in so doing substantially changed the composition of the oil so that its interfacial tension with water was reduced to $(\gamma_{o/w})_a$, where the subscript "a" stands for alcohol. As it turned out, $(\gamma_{o/w})_a$ asymptotically approaches 15 dynes/cm irrespective of the original value of $\gamma_{o/w}$. This made it possible to retain at least certain oil molecules in the interphase so that values of Π exceeded values of $(\gamma_{o/w})_a$.

This concept gave the formulator a ready tool to aid him in matching emulsifiers to oils for microemulsification. The trick, however, was in changing the composition of the oil to be emulsified with a minimum of cosurfactant and one which met the specifications of the job. This is discussed in greater detail in the next section.

One of the valuable results of this approach to microemulsion formulation has been the insight it has given to the specification of surfactants which are conducive to the formation of microemulsions. For any given surfactant, a short cosurfactant will promote a w/o system and a long cosurfactant an o/w system. In the case of soaps, the larger the size of the (hydrated) cation, the more effective will that particular soap be in promoting an o/w microemulsion. And finally, it was made clear that there is a specific structural inter relationship among surfactant, cosurfactant and oil molecules which promotes penetration of the oil into the interphase where it can then increase the value of Π. These are not inconsiderable contributions.

In concluding this discussion, it is only fair to say

that from an overall point of view, the concept of zero interfacial tension as a necessary prerequisite for microemulsion
stability is open to argument (cf. Chapter 5, VI,B; Chapters
6 and 7). The use of the film balance equation was an oversimplification. From the formulation perspective, however,
the expression $(\gamma_{o/w})_a$ is a valuable one, and its utilization
is recommended in the formulation context. On the practical
side, once an o/w system has successfully passed through the
viscoelastic gel stage and becomes a translucent o/w microemulsion, it remains stable for years. Whether this stability
is thermodynamic or kinetic is of small consequence.

IV. EMULSIFIABLE OILS

Before attacking the problem of choice of emulsifiers or
the employment of HLB or the other tools of the microemulsi-
fier, the formulator would be well advised to carefully exam-
ine the oil he is about to disperse. It is altogether fair to
say that this will not be the oil he will finally emulsify.
He will modify it unintentionally with cosurfactant or he may
deliberately dilute it with another oil to improve its emulsifiability or overall performance. In any event, a thorough
knowledge of each and every component of the oil will simplify
his task. In the case of natural products a search of the
literature is in order; with synthetics, infrared spectroscopy
and chromatography among other analyses are indicated. The
more one knows of the chemical makeup of his oil, the better
he will be in the long run. Two articles (10,11) indicate the
kind of study that is involved. One is concerned with why
borax is such a good emulsifier with Beeswax for making cold
cream, and the other tries to explain why Carnauba is an emulsifiable wax. These may not be the correct explanations, but
they illustrate the principles involved very well.
    In this perspective let us consider the effect of emulsi-
fier on the oil. By and large, microemulsions require two
kinds of emulsifiers, the primary surfactant (e.g., a soap)
and the cosurfactant or amphiphile, which may be deliberative-
ly added to the system or may be derived from the oil (cosurfactant *in situ*) or both. The primary surfactant usually distributes itself between the water and the interphase whereas
the cosurfactant partitions between the oil phase and interphase. Normally, almost all of the surfactant is in the interphase in a microemulsion system; a smaller fraction of the
total cosurfactant finds its way into the interphase, with the
balance remaining in the oil phase. It is this latter fraction which changes the composition of the oil, and so the original oil/water tension.
    Although it is difficult to measure the partitioning coef-
ficient of a given cosurfactant between oil and interphase,

there is an easy way to judge its efficiency as a microemulsi-
fier. It is simply to measure the interfacial tension of the
oil plus cosurfactant against water. Make a series of mixtures of oil and various percentages of cosurfactant up to
10%. Let us say that the cosurfactant is cetyl alcohol and
the oil a hydrocarbon having an interfacial tension against
water of 50 dynes/cm. If the tension of the mixture drops to
15 dynes/cm some place in the 10% range, a likely cosurfactant
candidate has been found. The lower percentage of cosurfactant needed to depress the tension to 15 dynes/cm, the better
the candidate. For these purposes the effect of temperature
is not too important since if it is necessary to heat the cosurfactant and oil to make them liquid, this will be the temperature at which the emulsification will take place anyway.

These measurements can be made on a routine basis with a
little ingenuity. A DuNouy tensiometer or a Wilhelmy plate
balance are well-established techniques to start with. The
latter is the preferred technique.

It is essential to recognize that partitioning is different from solubility of the materials. As a rule, amine soaps
are completely soluble in the oil phase plus cosurfactant.
Once water is added, however, the soap partitions between the
water phase and interphase, leaving negligible amounts left in
the oil phase. The cosurfactant partitions between the oil
phase and interphase but considerably more of it is left in
the oil phase than is primary surfactant in the water phase.

With further regard to the solubility of emulsifiers in
the oil phase, experience has been that most oils and emulsifiers are mutually soluble. The few exceptions are nonpolar
oils such as mineral oil or paraffin wax and alkali metal
soaps, particularly the laurates. Among the nonionics, the
organic nature of the polyoxyethylene chains tends to compati-
bilize these emulsifiers with most oils. The exceptions in
this case are the very highly ethoxylated emulsifiers which
are not compatible with nonpolar oils but may be compatible
with more polar ones.

Returning to the composition of emulsifiable oils, it is
now understandable why blending two or more oils with one another to achieve the desired polarity or emulsifiability is
such an attractive proposition. It is a feasible alternative
to the original approach so long as dilution or adulteration
of the primary oil is possible. It is a particularly good op-
tion when it is desirable to employ a fixed emulsifier system.
A good example of this was the self-polishing floor waxes
where low cost and good leveling properties were associated
with soap systems. There were a number of waxes both synthetic and natural which did not yield satisfactory microemulsions for self-polishing floor waxes until blended with other
waxes. A good way to lower $(\gamma_{o/w})_a$ was with Ouricury wax.

Although emulsions of this wax were too viscous on their own, blending with a paraffin or microcrytalline waxes made them into satisfactory (and cheaper) emulsions (12).

Ouricury wax contained a very high percentage of hydroxylated components, even higher than Carnauba. This hydroxyl value seemed to be the key to the emulsifiable waxes. It was by oxidizing microcrystalline, Fischer-tropsch, and polyethylene waxes, to mention just a few, that replacements for the exorbitantly expensive Carnauba wax were obtained. Oxidation (and esterification) was continued until just the right amount of alcohol, ester, and ketone was obtained to make a wax which produced a microemulsion with soap. In this way, cosurfactant was built into the wax.

The incidental production of aldehydes in these oxidized waxes degraded color and although these products performed well as emulsifiable waxes, they could not compete with the price, versatility, and color of the new (micro) emulsion polymers.* The result is that emulsifiable waxes are probably on the way out.

Of course, not all emulsifiable oils have built-in cosurfactants. Among these are turpentine (α-pinene), orthophenylphenol, many perfume oils (usually at very high levels of emulsifier), cyclohexane, kerosene, the normal paraffins from hexane to hexadecane, mineral oil, and Chlordane (octachlorindene), to mention a few.

By the terms of this concept microemulsions occur only with intermediately hydroxylated compounds. Below this range, $(\gamma_{o/w})_a$ has not been sufficiently depressed. Above this range, the predominantly alcoholic interphase squeezes oil molecules out of it or forms such a condensed film that the interphase does not bend sufficiently to yield small droplet sizes.

V. THE IMPASSE

It is the oils that are omitted from the above listings which eloquently tell why microemulsions are so rare. Benzene, toluene, and xylene can readily yield dispersions of the w/o type, but it requires a larger than normal amount of methylcyclohexanol and soap to make o/w microemulsions of them. In the case of the triglyceride, the situation is even more difficult. Twelve parts of Ouricury wax are needed to make eight parts of soya bean oil emulsifiable. Most other nonhydroxylated oils present similar problems; foremost among these are the paraffin and microcrystalline waxes.

---

*The editor wishes to thank Irving Y. Strauss of Dura Commodities Corporation for the background information he was kind enough to furnish in connection with emulsifiable waxes.

The answer lies in the presence of oil molecules in the microemulsion interphase and their interactions there with the other tenants. Any polar oil is surface active and will find its way into the interphase. Thus, a triglygeride is part of the interfacial film in the digestion of fats in the intestines (13). Unlike the molecules of nonpolar oils such as alkanes or benzene, which nestle among the hydrocarbon tails of conventional surface active tenants but never reach the water phase, the triglycerides are intercalated among such oriented tenants with their glyceride heads actually in the water phase. This has a profound effect on the behavior of the film, tending to expand it. The absence of proton donor groups on these glycerides greatly reduces the ability of such films to form microemulsions--they fail to remain coherent. The three ester linkages overwhelm any cohesional tendency that may be supplied by hydroxylic groups on adjacent tenants.

This is the root of the impasse in microemulsifying most oils. It is difficult to find surfactant and cosurfactant which will combine with them in the interphase to produce the state of film which is conducive to microemulsions formation. It is hoped that this delineation of the problem will serve as grist for the mill to increase the range of emulsifiable waxes.

VI. RHEOLOGY

The consistency of microemulsions is not as susceptible to alteration as is that of macroemulsions. Thickening agents which can control the rheological properties of conventional emulsions will more than likely destroy the stability of microemulsions. Accordingly, other means are utilized to vary their consistency.

Schulman and Cockbain (5) correlated the rheological properties of emulsions with the state of the film--the more solid the state of the film, the higher the consistency of the emulsion. Within limits, this applies rather elegantly to microemulsions. The use of more alcohol or low HLB nonionic as cosurfactant will increase the viscosity of the finished product and vice versa. This is a convenient and effective way to affect rheological properties.

The combination of microemulsions with liquid crystalline phase is another way to alter the viscosity of the finished emulsion. Since the lamellar or cylindrical phase are viscous and in thermodynamic equilibrium, this will also alter viscosity without destroying stability.

REFERENCES

1. Kai Li Ko (Yuan), Ph.D. Dissertation, Physics Department,

Tulane University, December, 1975.

2. Prince, L. M., *in* "The Chemistry and Manufacture of Cosmetics," (M. G. deNavarre, Ed.), Vol. III, pp 25-37, Continental Press, Orlando, Florida, 1975.

3. "The Atlas HLB System," Bulletin LD-97 3M 5/71, ICI United States, Inc., Wilmington, Delaware.

4. Becher, P., and Griffin, W. C., "HLB, An Explanation and Bibliography," *in* "Detergents and Emulsifiers," Allured Publishing Corporation, Ridgewood, New Jersey, 1974.

5. Schulman, J. H., and Cockbain, E. G., *Trans. Faraday Soc. 36*, 551 (1940).

6. Schulman, J. H., and McRoberts, T. S., *Trans. Faraday Soc. 42B*, 165 (1946).

7. Shinoda, K., *J. Colloid Interface Sci. 24*, 4 (1967).

8. Beerbower, A., and Hill, M. W., *in* "Detergents and Emulsifiers," Allured Publishing Corporation, Ridgewood, New Jersey, 1971.

9. Beerbower, A., and Hill, M. W., *Amer. Cosmetics Perfumery 87*, 85 (1972).

10. Prince, L. M., *Cosmetics and Perfumery 89*, 47 (1974).

11. Prince, L. M., *Soap Chemical Specialties 36*, Sept., Oct. (1960).

12. Prince, L. M., U.S. Patent 2,441,842 (1948); Warth, A. H., *in* "The Chemistry and Technology of Waxes," pp 723-724, 2nd ed., Reinhold, 1956.

13. Prince, L. M., *in* "Biological Horizons in Surface Science," p 361, Academic, 1973.

APPENDIX

Microemulsification--A Technical Explanation

The aim of this passage is to describe in simple terms
the molecular interactions which take place at the interface
between oil and water. It is these interactions which are responsible for the direction and degree of curvature of this
interface and, in turn, for the types of emulsions, i.e., w/o,
o/w, macro, or micro. These interactions are different from
those taking place in a bulk phase because they occur in two
dimensions instead of three. This imposes some restrictions
on their motions but is more than compensated for by a lowering of the interfacial tension between the mutually insoluble
liquids and thus greatly reduces the work needed to disperse
one liquid in the other.

An emulsion (a macroemulsion) was defined by Clayton as a
dispersion of two, or more, mutually insoluble liquids, one in
the other. The emphasis is on liquids. In most cases one of
the liquids is water. The other is a water-insoluble liquid--
an oil, of animal, vegetable, mineral, or synthetic origin.
This oil must be liquid at the temperature of emulsification.
If it is not, steps must be taken to make it so. When, after
emulsification, this oil freezes or congeals, the system is
technically no longer an emulsion but a dispersion. It is
common practice, however, to call such systems emulsions because they are made by an emulsification process. Margarine is
is such a dispersion. Its properties, therefore, depend upon
the way it is emulsified and how it is cooled.

Because the emulsification process involves only liquids,
the surface tension forces at the boundary between the two immiscible fluids are free to exert their influence equally in
all directions. Thus, the dispersed phase assumes the form of
spheres since this is the geometric shape which possesses the
minimum surface area per unit volume. These spheres may consist of water or oil. When they are water, the emulsions are
called water-in-oil, w/o; when they are oil, the emulsions are
called oil-in-water, o/w. Common examples of w/o emulsions,
aside from margarine, are cold cream, hydraulic fluids, printing inks, and dry cleaning fluids. Examples of o/w emulsions
are asphalt, floor polish, pharmaceutical preparations, jet
fuels, latex for paint, mayonnaise, ice cream and milk, to
mention only a few.

All of these systems have one thing in common. The droplets of the dispersed phase are very small, ranging in diameter

51

from one thousandth to one millionth of a centimeter. This
produces an enormous increase in the area of contact between
the two liquids. The effect upon the surface area per gram of
dispersed phase as one bulk liquid is broken up into smaller
and smaller droplets is illustrated in Table 1. To associate
these dimensions with a familiar scale, they are listed in
several units. The magnitude of the areas can be appreciated
when it is considered that the dispersion of 700 grams of water into droplets 1 micron in diameter creates more than an
acre of interfacial area.

Such comminution of a bulk liquid into droplets of microscopic size is effected by doing mechanical work on the
fluids. In margarine systems, for example, the mechanical
process of emulsification takes many forms. They are summed
up by the thermodynamic equation

$$Work = -\gamma_i dA$$

where γ_i is the interfacial tension between the two immiscible
liquids when emulsifying agent is present in the system and dA
is the increase in the area of contact between the two liquids.

This equation identifies interfacial tension as a thermodynamic function and relates it to emulsion properties. For
example, when γ_i is low, of the order of magnitude of a fraction of a dyne/cm, emulsions are of smaller droplet size and
usually more stable than when the tension is higher. It is
also clear that a decrease in the value of γ_i decreases the
work needed for emulsification. When the interfacial tension
is zero, as it presumably is in microemulsions, the work term
becomes zero and emulsification occurs spontaneously. When
mechanical work is required to form these dispersions of liquids, Bancroft called them (macro) emulsions; when the work
done is of chemical origin, Schulman called the systems microemulsions.

Of importance to our molecular thesis is the fact that
as a thermodynamic function, γ represents the summation of the
forces of interaction among the molecules at or near the interface. It represents the summation of these forces in two
dimensions just as pressure, another thermodynamic function,
is the summation of the forces exerted in three dimensions by,
for instance, exploding gasoline molecules on the piston of an
internal combustion engine. It is in this frame of reference
that we approach the concept of interfacial tension in emulsion systems.

To assume our molecular-thermodynamic stance, let us begin by looking at the values of the interfacial tension
against water of a number of liquids that are frequently emulsified with water (Table 2). We shall use the term $\gamma_{o/w}$ to
denote these values. When there are no emulsifiers in the

system $\gamma_{o/w} = \gamma_i$. The data indicate that $\gamma_{o/w}$ depends upon the chemical constitution of the oil and that to lower it, it is necessary to make the oil more like water. Double bonds and polar groups serve this purpose.

TABLE 1

Droplet Diameter vs. Surface Area

A	cm	microns	meters^2/gram
100,000	0.001	10	0.6
10,000	0.0001	1	6
1,000	0.00001	0.1	60
100	0.000001	0.01	600

But since we do not normally wish to change the composition of the liquids we are about to emulsify, the next best thing to do in order to lower the interfacial tension is to introduce into the system materials which, at low concentration, alter the boundary between the immiscible liquids. Such materials are called emulsifying agents. They occur in the form of finely divided solid particles or discrete molecules. In either case their mode of operation is the same. They spontaneously interpose themselves between the two liquids in a layer that is one powder particle or one molecule thick. Our present interest is only in those agents which are easily dispersed as discrete molecules.

TABLE 2

Interfacial Tension against Water

20°C

Oil	$\gamma_{o/w}$, *dynes/cm*
Mineral oil	55
Carbon tetrachloride	45
Benzene	35
Cottonseed oil	30
Nitrobenzene	25.7
Ethyl ether	10.7
n-Octyl alcohol	8.5
n-Butyl alcohol	1.8

The way in which these molecules orient themselves after they reach the interface is extraordinary. This is ascribed to the fact that a portion of such molecules are strongly attracted to oil and another portion is strongly attracted to water. The hydrophile-lipophile balance, or HLB, is a convenient measure of these counter attractions or solubilities. From the viewpoint of the emulsifier molecule, this causes it to be pulled in one direction by oil molecules that want it to dissolve in their phase and in the other direction by water molecules that want it to enter their phase. This tug-of-war for its body adlineates each emulsifier molecule perpendicular to the interface and parallel to one another. This is quite different from the random orientation of molecules in a bulk liquid phase. This ordered arrangement of molecules is called a monomolecular film or monolayer. It is responsible for most of the properties of emulsions both macro and micro.

A schematic diagram of a wedge of the monolayer of a dodecane-in-water microemulsion is shown in Fig. 1. This emulsion is stabilized with oleate soap and cetyl alcohol, cf. Formula A above. In this diagram the molecules of the emulsifying agent are represented in the conventional manner, more or less to scale. The zigzag lines represent the aliphatic tails of the surfactant (soap) and cosurfactant (cetyl alcohol) as well as whole nonpolar hydrocarbon molecules. The circles attached to these tails represent the polar heads or water soluble portions. The oleate tails are 25 A long and each tail occupies a cross sectional area of about 20 A^2. If one considers the tails to be square in cross section, they are 4.5 A on a side. The circles by themselves are water molecules and are 2.7 A in diameter. Circles with plus signs in them represent the cations. The hydrocarbon tails of the emulsifiers (surfactant and cosurfactant) dissolve in the oil phase while the polar heads are buried in the water phase. To the best of our current knowledge this is what a microemulsion interface looks like.

The next question is how do molecules which are oriented in this way interact with one another. First of all, the attractive and repulsive forces among the heads are different than among the tails. Among the tails, London Dispersion forces are inversely proportional to the seventh power of the lateral distance between them. Thus, it is not until a high concentration of molecules exists in the interface and the tails and hydrocarbon molecules begin to crowd one another that they develop mutual repulsion. When this occurs, the lateral pressure develops very abruptly because of the seventh power relationship. On the other hand, the forces operative among the heads mainly depend upon hydrogen bonding with each other and water molecules. The lateral forces here are more dependent upon composition than crowding. The significant point is that as the concentration of emulsifier in the interface increases, the net repulsion among the heads and tails increases and rapidly develops a two-dimensional pressure. As

noted, this can attain a value as high as 15 dynes/cm. Observe that this pressure is expressed in dynes/cm and not
dynes/cm^2. An idea of the magnitude of this two-dimensional
pressure can be obtained by dividing it by the film thickness
of 25 A or 25 x 10^{-8} cm. This is a three-dimensional pressure
of 60 million dynes/cm^2. Since 1 atmosphere equals a million
dynes/cm^2, this is 60 atmospheres or 900 lb/in.^{2}, a not inconsiderable pressure even in our macroscopic world.

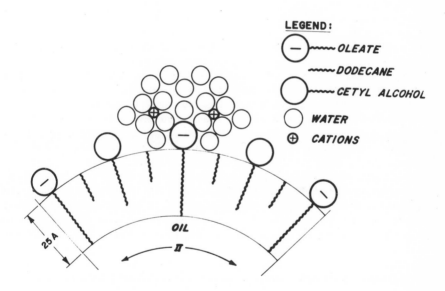

*Fig. 1. Schematic diagram of dodecane-in-water micro-
emulsion stabilized with oleate soap and cetyl alcohol.
The positions of and interrelationships among the several species in and near the interphase are shown in
accordance with theory.*

Such a picture enables us to determine in an elementary
way how the collective interactions among the highly oriented

interphase species control the direction and magnitude of its
curvature. This is well illustrated in Fig. 3 of Chapter 5,
although the oil molecules are not shown in this interphase.

By using this schematic diagram, the question can be an-
swered as to why one liquid disperses in the other instead of
vice versa. Consider that the interphase is capable of possessing different tensions at each of its sides--a reasonable
assumption since it has already been indicated that the forces
among the heads are different from those among the tails.
Thus, in the flat film prior to curvature, if the pressure at
the oil side is greater than at the water side, the film will
bend so as to envelop water in droplet form. As the film
bends under the influence of the molecular forces at play, the
pressure at the oil side is relieved so that after curvature,
the pressure or tension at each side is the same. When an o/w
emulsion is desired, it is only necessary to increase the repulsive forces among the heads of the interphase tenants.
This is done by increasing the effective size or number (or
both) of the heads of ionic emulsifiers or the number of the
ethylene oxide groups on nonionic emulsifiers. We call this
raising the HLB. This makes the two-dimensional pressure at
the water side greater than that at the oil side so that the
curved film envelops oil in droplet form.

As for the magnitude of film curvature, this, too, is dependent upon the molecular interactions among film tenants.
The equilibrium or stable droplet size depends upon the initial two-dimensional pressure gradient across the flat interphase, i.e., upon the ratio of the pressure at the oil side to
that at the water side. This, in turn, depends upon the chemical nature of the tenants.

In conclusion, it seems important to point out that the
volume of the interphase of a microemulsion is a very appreci-
able percentage of the total volume of the dispersed droplet
(core + interphase). These geometric relationships are shown
in Table 3. They emphasize the magnitude of the role played
by the interphase in the formation of microemulsions.

TABLE 3

Interphase Percentages in Microemulsions[1]

Total Droplet Diameter[2]		Interphase Volume
A	micron	%
1000	0.1	14
750	0.075	19
500	0.05	27
250	0.025	49
100	0.01	88

1. For 25 A thick interphase.
2. Core + interphase.

How to Formulate Microemulsions
with Less Surfactant

KOZO SHINODA AND HIRONOBU KUNIEDA

Department of Chemistry
Faculty of Engineering
Yokohama National University
Yokohama, Japan

I. INTRODUCTION

Critical opalescence arises from large fluctuations of concentration, density, or refractive index. It occurs above the critical temperature of complete miscibility of liquids · and is due to the association of individual molecules into aggregates of colloid dimensions. The turbidity, i.e., the total relative amount of light scattered, is large, because the fluctuation of concentration, density, or refractive index is large. This phenomenon is observed, however, in a very narrow range of temperature, close to the critical composition so that it is hard to utilize such an optical tool with products such as pharmaceuticals, foods, and cosmetics. On the other hand, micellar solutions intrinsically possess concentration, density, and refractive index fluctuations. A micellar solution can therefore scatter light at ordinary temperatures and concentrations. If the size of micelle becomes larger by the addition of cosurfactant and/or by adding oil (or water), its turbidity increases and the solution appears bluish-white or white--indicating that it is a microemulsion. Since swollen micellar solutions* look turbid but are infinitely stable over wide concentration and temperature ranges, the phenomenon of normal opalescence is quite prevalent. Polymer solutions may also be utilized to formulate microemulsions. Section II reviews the optical properties of microemulsions from a basic standpoint.

---

*If a system consists of a single liquid phase, we shall call it a solution regardless of the state of solution. If a system consists of more than one phase, we shall call it a system, emulsion or two phase solution.

Microemulsions are prepared in oil-continuous as well as water-continuous media. The relationship between both types of microemulsions is clearly understood by the studies of phase diagrams composed of (1) water, oil, and surfactant as a function of temperature and (2) water, oil, surfactant, and cosurfactant at a given temperature. The change of temperature corresponds to a change in the HLB of the surfactants. The phase inversion temperature (PIT) is observed between the optimum temperature for w/o type and o/w type microemulsions. Hence, PIT (HLB-temperature) is a very important characteristic property of surfactant solutions for understanding emulsions, solubilization, and microemulsions. Finally, the reason "why w/o type microemulsions are easier to prepare than the o/w type" is explained.

Section III serves to explain microemulsions comprehensively.

In Section IV, ways to formulate microemulsions with less solubilizer (emulsifier) are listed and explained. The behavior of nonionic surfactant systems markedly vary with temperature changes, for example, from o/w to w/o type. This change is very useful for some purposes but not for others. Properties of solutions (systems) containing ionic surfactant are not affected as much by temperature change. Hence, both nonionic and ionic surfactants are useful, depending on the purpose for which they are used. The last section is useful for selecting suitable surfactants to formulate microemulsions.

II. OPTICAL IDENTIFICATION OF MICROEMULSIONS

A. Critical Opalescence

Close to any critical point, large fluctuations of concentration, density, refractive index, etc., occur, because the free energy changes very slightly with composition and allows the fluctuation to become large. If these fluctuations are accompanied by changes in the refractive index, the system is no longer optically uniform and the amount of light scattered will be large.

Lord Rayleigh (1) derived an equation for this kind of opalescence. The turbidity τ, i.e., the total relative amount of light scattered by a unit volume of the substance in all directions is defined by

$$\frac{I_o - I_{sc}^{total}}{I_o} = e^{-\tau} \qquad \ldots \ (1)$$

and if τ is small

$$\tau = \frac{I_{sc}^{total}}{I_o} = \frac{32\pi^3 n^2 \overline{(\delta n^2)} \Delta V}{3\lambda^4} \qquad \cdots \text{(2)}$$

where τ is turbidity, λ the wavelength of light, I_o the intensity of the original incident beam, I_{sc}^{total} the loss of intensity of the original incident beam, n the refractive index of medium, ΔV the volume of the scattering unit, and $\overline{(\delta n^2)}$ the mean square fluctuation of n within ΔV.

Expressions for critical opalescence were independently offered by Einstein (2) and Smoluchowski (3). In binary solution $\overline{(\delta n^2)}$ was defined as follows:

$$\overline{(\delta n^2)} = \left(\frac{\delta n}{\delta\rho}\right)^2 \overline{(\delta\rho^2)} + \left(\frac{\partial n}{\partial c}\right)^2 \overline{(\delta c^2)} \qquad \cdots \text{(3)}$$

where ρ is the density, $\overline{(\delta\rho^2)}$ the mean square fluctuation of ρ, c the concentration of solute 2 in grams per cubic centimeter, and $\overline{(\delta c^2)}$ the mean square of concentration fluctuation. The relation between $\overline{(\delta c^2)}$ and G, the free energy of component 2 is

$$\overline{(\delta c^2)} = \frac{kT}{\Delta V(\partial^2 G/\partial c^2)} \qquad \cdots \text{(4)}$$

$\overline{(\partial c^2)}$ in Equation (4) is large, because the change of free energy with concentration is so small at the critical point. On the other hand, the first term on the right-hand side of Equation (3) does not change much in a binary solution. Substituting Equations (4) and (3) into (2), we obtain

$$\tau = \frac{32\pi^3 n^2}{3\lambda^4}\left(\frac{\partial n}{\partial c}\right)^2 \frac{kT}{\partial^2 G/\partial c^2} \qquad \cdots \text{(5)}$$

The solution is turbid close to the critical point; however this scattering diminishes rapidly, both with rising temperature and with changing composition, making it difficult to utilize for all practical purposes. Table 1 illustrates scattering data for the system $CCl_4 + C_6F_{11}CF_3$ investigated by Zimm (4).

B. Light Scattering of Micellar Solutions

A micelle is a stable colloidal particle existing in a solution which possesses concentration fluctuation within

TABLE 1

Scattering of Light Close to the Critical Temperature

t, °C	Turbidity*
28.31	>20
28.33	7
28.34	4.0
28.41	1.13
28.50	0.56
28.90	0.150
29.80	0.052
34.7	0.0089
47.2	0.0025

*Turbidity, τ, is indicated by Equation (1). Phase separation does occur at 28.31°C in this system.

itself. The partial molal free energy of a micellar surfactant does not change appreciably with concentration (5,6). Concentration fluctuation of micelles also occurs.* For example, a micellar solution can scatter light at ordinary temperature and concentrations, but the turbidity of the solution is small. Rayleigh (7) derived an equation for the turbidity of a particulate suspension in which

$$\tau = \frac{32\pi^3 n^2 \Delta n^2 N V^2}{3\lambda^4} \qquad \cdots \ (6)$$

where τ is the turbidity, λ the wavelength of light, Δn the difference of refractive index between medium and scattering unit, N the number of scattering units per unit volume, and V the volume of the scattering unit. In order to obtain so-called microemulsions, i.e., to increase the scattering of light, the size of micelle, V, in Eq. (6), has to be large. For this purpose the addition of cosurfactant and solubilizate in the case of ionic surfactants, or the adjustment to optimum temperature in the case of nonionics, is necessary.

1. Ionic Surfactants

Schulman and co-workers obtained a faintly turbid solution

---

*This fluctuation is not taken into account in the determination of micellar size by light scattering.

upon adding alcohol to emulsions composed of water, hydrocarbon and ionic surfactant. They called this solution a microemulsion. We infer from this name that the diameters of the emulsion droplets are very small owing to the presence of an appropriate amount of surfactant. Actually, Schulman and his co-workers consider this system not as a micelle but as an emulsion (8,9,10). Unlike ordinary emulsions, however, they appear to be absolutely stable towards phase separation. Extensive investigation by Schulman and his co-workers by low-angle X-ray scattering (11), light scattering (12,13), ultracentrifuge (14), electron microscopy (8), NMR (13,15), etc., led them to consider that colloidal particles of 100-1000 A diameter were present. Particles in the lower range are comparable with the size of micelles (16). Schulman reported several compositions which yielded microemulsions (14). Later studies (17,18) of the phase diagram of one of them, namely, water + benzene containing 25 wt%/system of surfactant (potassium oleate + hexanol or pentanol) revealed that Schulman's microemulsions (14) belong to a solubilized oil-continuous micellar realm as shown in Fig. 1. In this figure, the weight fraction of solvent (oil + water) 75 wt%/system is plotted horizontally and that of solubilizer 25 wt%/system is plotted vertically so that the left axis of ordinates represents 75 wt% water + 25 wt% solubilizer and vice versa. Realms I$_o$ and II are oil continuous single-phase and two-phase solutions, respectively. The turbidity of the solution continuously increases with the increase of the amount of solubilizate, because the scattering of light from micellar solution depends on the size, number, and the difference in refractive index of swollen micelles. When the solution is not too turbid, the solution looks bluish-white if observed perpendicular to incident light and looks yellowish if observed in transmitted light. The transitions to ordinary emulsions that occur when excess water is added to the system appear in no fundamental way different from those that occur in ordinary micellar solutions. Hence, Schulman's microemulsion is considered a solubilized micellar solution (17-26). Large swollen micelles are responsible for the turbidity of the solution.

2. Nonionic Surfactants

An aqueous solution of nonionic surfactant splits into two phases above the cloud point. Thus a cloud point curve is a liquid-liquid solubility curve with a lower consolute temperature (27,28). Micelles become very large close to the cloud point particularly in the presence of solubilizate (29,30,31). An aqueous solution of (5-45 wt%) nonionic surfactant close to the cloud point looks bluish-white, i.e., exhibits opalescence. The addition of solubilizable oil such as cyclohexane swells the micelles and the solution becomes densely turbid (18).

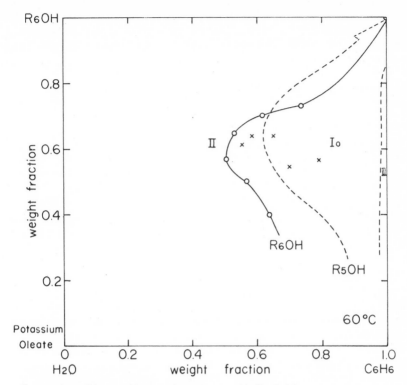

Potassium Oleate+Hexanol: 25wt%, H2O+C6H6: 75wt%.

*Fig. 1. The phase diagram of water + benzene con-
taining 25 wt%/system of potassium oleate + hexanol
(or pentanol) at 60°C. The weight fraction of solvent 75 wt%/system is plotted horizontally and that
of solubilizer 25 wt%/system is plotted vertically.
Schulman's microemulsions, indicated by x, belong to
an oil-continuous single-phase realm. Reproduced
from Shinoda and Kunieda, Ref. (18), courtesy of
Academic Press, Inc.*

Optimum temperature for large scattering of light in nonionic
surfactant systems can be adjusted by adding suitable amounts
of either hydrophilic or lipophilic nonionic surfactants.

C. Solutions of Proteins or Water-Soluble Polymers

    Water-soluble polymers consisting of hydrophilic and hy-
drophobic groups dissolve in water, orienting hydrophilic
groups towards water and hydrophobic groups inwards. This
phenomenon is similar to micellar dispersion. The cloud point

is observed in these nonionic water-soluble polymer solutions
at higher temperature (32,33,34). The partial molal Gibbs
free energy is nearly constant over a wide concentration range
(35). Strauss and Williams (36) have studied light scattering
from aqueous poly-soap solutions. Scattering increases as
solubilizate is added. Such polymer solutions may also be
utilized to prepare microemulsions.

III.   RELATIONSHIP BETWEEN W/O AND O/W MICROEMULSIONS

A.   The Effect of Temperature on the Phase Equilibria and the
     Types of Dispersions in a Ternary Nonionic System

     A swollen micellar solution of nonionic surfactants close
to the cloud point is identical to a Schulman microemulsion.
Thus, the study of phase equilibria in ternary systems composed of water, hydrocarbon, and nonionic surfactant is important in the preparation of microemulsions and in increasing
the mutual dissolution of water and oil by the action of a
surfactant.   There are many variables in this study, such as
the types of oils, the kinds of surfactants, the composition
of components, the effect of additives, and temperature.
Among saturated hydrocarbons, cyclohexane may well represent
the typical behavior of an oil in these systems.   The phase
diagram is nearly symmetrical if cyclohexane is used as the
oil phase.   Although the composition of water vs. oil has to
be varied over the entire volume fraction range, the concentration of the surfactant may be fixed at 1% to 10% from a
practical viewpoint.   According to Schulman's recipes (14),
20-40 wt% of surfactants were necessary but 5-10 wt% of surfactants is sufficient to produce microemulsion as described
below.   Although little attention has been paid to the effect
of temperature on the solubilization, emulsion types, and the
dissolution state of an anionic surfactant, these effects have
to be thoroughly explored in systems containing nonionic surfactants because the effects are so remarkable and important.
Phase equilibria and dispersion types of water-cyclohexane
systems containing 7 wt% per system of $i\text{-}C_9H_{19}C_6H_4O(CH_2CH_2O)\text{-}$
$_{9.7}H$ as a function of temperature have been examined (37).
The study of these systems has been useful in understanding
the mutual relations between (a) solubilization of oil in
aqueous surfactant solutions (38); (b) solubilization of water
in nonaqueous surfactant solutions (39); (c) the types, inversion, and stability of emulsions (40); and (d) practical applications, such as washing, dry-cleaning, and emulsification.
     The phase diagram of a water-cyclohexane system containing
7 wt% polyoxyethylene (9.7) nonylphenylether is shown in Fig.
2.   The left-hand side of the figure corresponds to an aqueous

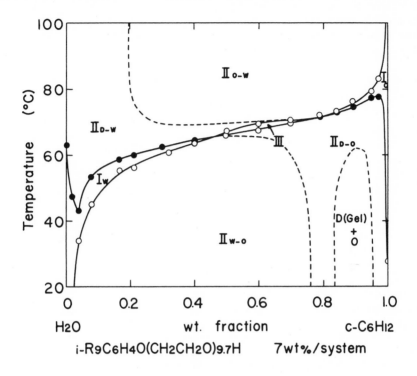

*Fig. 2. The phase diagram of the water-cyclohexane
system containing 7 wt% of polyoxyethylene (9.7) nonylphenylether as a function of temperature. Cloud point
curve is indicated by ● in I_w region and haze point
curve, by ○ in I_O region. Reproduced from Shinoda and
Saito, Ref. (37), courtesy of Academic Press, Inc.*

surfactant solution containing a small amount of cyclohexane.
Realm I_w is the oil-swollen micellar solution. The solubilization curve as a function of temperature is observed at a
relatively low temperature. Solubilization of cyclohexane in
an aqueous surfactant solution increases markedly close to the
cloud point, but above the cloud point a surfactant phase
separates from water and there is no solubilization in the
aqueous solution. A large amount of water and cyclohexane
dissolves in the surfactant phase, and the two phases (water
and surfactant phases) coexist above the cloud point curve,
realm II_{d-w}. If the amount of oil in the system is increased
at this temperature, an oil phase appears. The central realm
indicated by III describes a three-phase region composed of
water, surfactant, and oil phases. The importance of the

surfactant phase has been emphasized by Lapczynska and Friberg (41). When the volume fraction of the surfactant phase is large (about 80% in 7 wt% solution), the water or oil phase will disappear, as a result of a small change of composition or temperature. Thus, the three-phase realm is narrow and small. It becomes larger in more dilute solution. Realm II_{d-o} is a two-phase region consisting of surfactant and oil phases. Since with increasing temperature the solubility of water in a surfactant phase decreases, and that of oil increases, the result is an increase of volume of the water phase and a decrease of surfactant plus oil phases.

The right-hand side of Fig. 2 corresponds to a nonaqueous solution of a nonionic surfactant containing a small amount of water. Realm I_o is a water-swollen micellar solution of cyclohexane. The solubilization curve of water in cyclohexane is observed at a relatively high temperature. Solubilization of water increases as temperature decreases, particularly near the cloud (haze) point in a nonaqueous surfactant solution but a surfactant phase separates from cyclohexane below the cloud (haze) point. Above the I_w, III, and I_o regions two phases exist. Since the solubility of the surfactant in water is very small in this region (or in this temperature range), the aqueous phase is nearly pure water. The concentration of the surfactant in the oil phase increases with the change of composition from the right-hand side to the left-hand side, and, finally, the surfactant becomes continuous in the nonaqueous region. This tendency is strong at lower temperatures (below the haze point). The right-hand side of the dotted line indicates the two-phase solution consisting of water and oil phases, and the left-hand side of the dotted line indicates the two-phase solution consisting of water and surfactant phases. Similarly, two phases coexist below the I_w, III, and I_o regions. The surfactant dissolves in water at this temperature but does not dissolve well in oil. At the left-hand side of the two-phase region, excess oil separates from the oil-swollen micellar solution, realm II_{w-o}. However, the relative concentraion of the surfactant in the water phase increases with the change of composition from left to right, and, finally, the surfactant phase becomes continuous. Hence, the two phases on the right-hand side of the dotted line consist of surfactant (either liquid crystal or sol) and oil phases. The change in either hydrophilic or lipophilic chain length of the surfactant shifts the phase equilibria and dispersion types either to higher or lower temperatures, but the pattern is similar. These situations were envisaged in the PIT vs. phase-volume curves as a function of the hydrophilic chain length of nonionic surfactant as illustrated in Fig. 3 (42).

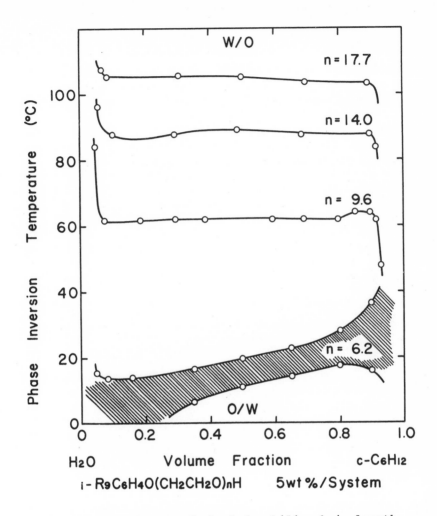

Fig. 3. *The effects of the hydrophilic chain length of nonionic surfactants on the PIT vs. phase-volume curves of cyclohexane-water emulsions. N is the mean oxyethylene chain length. Reproduced from Shinoda and Arai, Ref. (42), courtesy of Academic Press, Inc.*

B. The Effect of the Nonionic Oxyethylene Chain Length Distribution on the Phase Equilibria Diagram

It is evident from Fig. 2 that the solubilization in non-ionic surfactant solution markedly increases at the optimum temperature (close to the PIT). If the oxyethylene chain length of nonionics is made longer or shorter, a similar phase diagram is obtained, shifted to higher or lower temperatures, respectively, as shown in Figs. 2 and 4. If the temperature of the system is raised, the interaction between water and the hydrophilic moiety of the surfactant decreases. Thus, the effect of temperature increase and the decrease in the oxyethylene chain length in the surfactant molecule, may be similar.

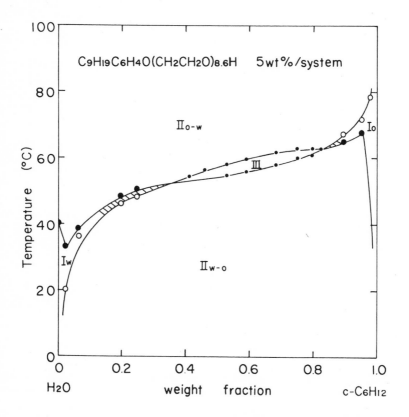

Fig. 4. The effect of temperature on the phase diagram of water + cyclohexane containing 5 wt%/system of $C_9H_{19}C_6H_4 O(CH_2CH_2O)_{8.6}H$. *Reproduced from Shinoda and Kunieda, Ref. (18), courtesy of Academic Press, Inc.*

This reasoning is confirmed by the phase diagram of nonionic surfactants in H_2O + c-C_6H_{12} as a function of the ethylene oxide chain length of surfactant as shown in Figs. 5 and 6. Here the oxyethylene chain length of surfactant in the ordinate decreases, instead of temperature, as in Fig. 4. We may conclude from Figs. 5 and 6 that the area of the so-called

microemulsion realm increases, provided the distribution of
the oxyethlyene chain length of solubilizer becomes narrower.
This conclusion is also supported by Lapczynska and Friberg
(41).

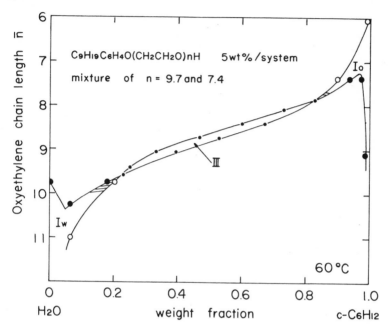

*Fig. 5. The effect of the average oxyethylene chain
length of nonionics on the phase diagram of water +
cyclohexane containing 5 wt%/system of the mixture of
$C_9H_{19}C_6H_4O(CH_2CH_2O)_{7.4}H$ and $C_9H_{19}C_6H_4O(CH_2CH_2O)_{9.7}H$.
The temperature increase in Fig. 4 and the decrease
of oxyethylene chain exhibit practically the same
effect. Reproduced from Shinoda and Kunieda, Ref.
(18), courtesy of Academic Press, Inc.*

The effect of temperature was significant in solutions of
nonionic surfactants, but not in solutions containing ionic
surfactants. It is now evident that the effect of temperature
increase in solutions of nonionic surfactants is equivalent to
increasing the fraction of lipophilic surfactant in solutions
containing in ionic surfactant. If so, the same phenomena as
in Figs. 5 and 6 may be observed in solutions containing
ionic surfactants.

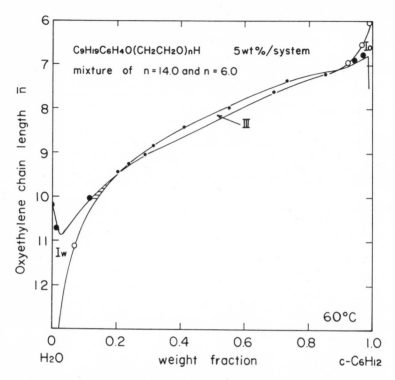

Fig. 6. *The effect of the average oxyethylene chain
length of nonionics on the phase diagram of water +
cyclohexane containing 5 wt%/system of the mixture
of $C_9H_{19}C_6H_4O(CH_2CH_2O)_{6.0}H$ and $C_9H_{19}C_6H_4O(CH_2CH_2O)_{14.0}H$.
The solubilized regions are diminished compared with
Fig. 5. Reproduced from Shinoda and Kunieda, Ref.
(18), courtesy of Academic Press, Inc.*

C. The Effect of the Ratio of Ionic Surfactant to Co-
    Surfactant on the Mutual Solubilization of Water
    and Oil

A lipophilic cosurfactant, such as alcohol, greatly assists in the formation of microemulsions in solutions of ionic
surfactants because ionic surfactants are usually too hydrophilic. The optimum ratio of ionic surfactant to cosurfactant is an important factor in enhancing solubilization just
as is the distribution in solution of the mixed nonionic surfactants shown in Fig. 5.

The phase diagram of a system which contains a total of
20 wt% of octylamine and octyl ammonium chloride and a total
of 80 wt% of water and p-dimethylbenzene are shown in Fig. 7
(43). Three liquid phases, L_1, L_1', L_2, and one liquid

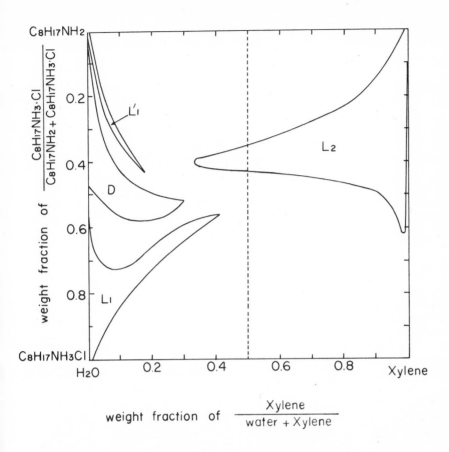

Fig. 7. Phase equilibria of octylamine, octyl ammonium chloride, water, and p-xylene at 22°C. Octylamine + octyl ammonium chloride, and p-xylene + water were kept 20 wt% and 80 wt% of the system, respectively. Reproduced from Ahmad, Shinoda, and Friberg, Ref. (43), courtesy of Academic Press, inc.

crystalline phase, D, were found. The region L_2, a nonaqueous phase, was formed by amine dissolved in water and p-xylene. The region L_2 was not found to extend to the p-xylene axis, because octyl ammonium chloride did not dissolve in p-xylene. The small amount of water hydrated the surfactant and cosurfactant, and the solubilization of water increased in the L_2 region. Maximum solubilization of water was found at an amine/ammonium chloride ratio of 1.5.

In the region L_1, the octyl ammonium chloride dissolved in water and solubilized p-xylene. Maximum solubilization extended up to an amine/ammonium chloride ratio of 0.8 (w/w).

Phase separation started above this ratio, and a neat phase, D, with a lamellar structure appeared. The solubilization of water by the neat phase became maximum when the ratio between amine and ammonium chloride was 1.0. On the water- and octylamine-rich corner, a phase, L_1', was found to occur beyond the neat phase, D. Octylamine dissolved in water and solubilized 16% p-xylene at an amine/ammonium chloride ratio of 1.3.

This type of phase diagram is convenient for determining the optimum mixing ratio of the two surfactants and also for determining the extent of solubilization of oil in water and water in oil in the presence of a definite amount of total surfactant (44). Earlier published triangular phase diagrams by Friberg (41) of water, p-xylene, and various mixing ratio of R_8NH_2 and R_8NH_3Cl give essentially the same information (45).

Octylamine may be changed to octylammonium hydroxide in a four-component system. Octylammonium hydroxide may be a reasonably well balanced surfactant and may not be very soluble in hydrocarbon. This may be the reason that there existed both an oil swollen aqueous micellar solution and a water swollen nonaqueous micellar solution. If alcohol had been used as a cosurfactant, which is readily soluble in xylene, a nonaqueous micellar solution region would not have appeared.

In order to illustrate the effect of the types of gegenions on ionic surfactants and the effect of the HLB of the cosurfactant, the solubilization of cyclohexane in aqueous metal dodecylsulfate-cosurfactant solutions is shown in Fig. 8. The weight fraction of solubilizer (metal dodecylsulfate + cosurfactant) 10 wt%/system is plotted horizontally and that of solvent(c-C_6H_{12} + water), 90 wt%/system is plotted vertically. The solubilization is maximum at the optimum ratio of ionic surfactant to alcohol ($R_8OH/R_{12}SO_4Na$; 0.7, $R_8OH/R_{12}SO_4\frac{1}{2}Mg$; 0.3, $R_8OC_2H_4OH/R_{12}SO_4\frac{1}{2}Mg$; 0.6 (w/w)). Excess oil separates beyond the solubilization curve up to the optimum ratio and surfactant phase separates if excess alcohol is added. The maximum solubilization at the optimum ratio and surfactant phase separation is very similar to the phenomena observed in the aqueous solution of surfactants mixtures in Fig. 5. The solubilization of cyclohexane in metal dodecylsulfate solution without cosurfactant is small. This may mean the hydrophilic property of this ionic surfactant is too strong and the size of micelle is small. The solubilizing power of magnesium dodecylsulfate is larger than that of sodium salts. This means the types of gegenion (cation) affects the HLB of ionic surfactants, and magnesium salt is a better balanced one. Actually, the optimum ratio of surfactant to octanol at which the hydrophile-lipophile property just balances, is much wider in the case of the magnesium salt. The solubilizing power of the surfactant mixture is larger, the closer the

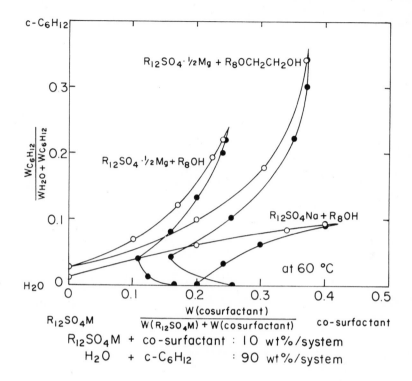

Fig. 8. The phase diagram of water + cyclohexane containing 10 wt%/system of metal dodeyl sulfate + octanol (or monoethyleneglycol octyl ether) at 60°C. The weight fraction of solvent 90 wt%/system is plotted vertically and that of solubilizer 10 wt%/system is plotted horizontally. Reproduced from Shinoda, Ref. (52), courtesy Academic Press, Inc.

HLB of ionic (hydrophilic) surfactant is to that of the lipophilic cosurfactant. The same conclusion may be drawn for the solubilization of water in a nonaqueous surfactant solution. But the amount of oil soluble ionic surfactant, such as dimethylethanol ammonium oleate (AMP-oleate)* which Schulman also tried, or Aerosol OT seems well balanced because the solubilization of water in the absence of cosurfactant is large and the solubilization attains a maximum with a small amount of added cosurfactant (Fig. 9).

---

*The correct name of dimethylethanolamine is 2-amino-2-methyl-1-propanol (AMP).

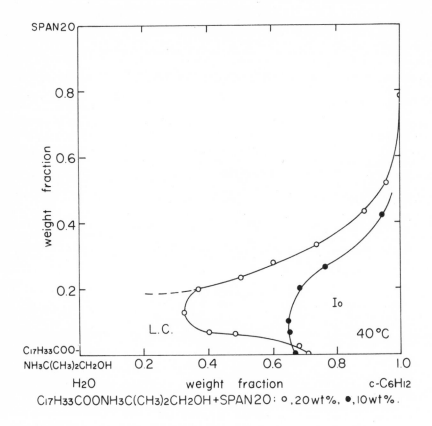

Fig. 9. The phase diagram of water + cyclohexane containing 10 and 20 wt%/system of dimethylanolamine (AMP) oleate ($C_{17}H_{33}COONH_3C(CH_3)_2CH_2OH$) + Span 20 at 40°C: (o) 20 wt% and (•) 10 wt%/system of solubilizer. Reproduced from Shinoda and Kunieda, Ref. (18), courtesy of Academic Press, Inc.

D. The Importance of the Phase Inversion Temperature
    (HLB-Temperature) and the Oxyethylene Chain Length
    Distribution of Surfactant (HLB-Ratio)

The important feature of solutions of a nonionic surfactant is the notable increase in the solubility of oil in an aqueous surfactant solution at the cloud point and of water in a non-aqueous surfactant solution at the haze point as shown in Figs. 2 and 4 (40). This phenomenon is also exhibited between these two extremes by the fact that a large amount of oil and water dissolves into the surfactant phase as shown in Fig. 10 (47).

Figure 10 illustrates the change of volume fractions of water, oil and surfactant phases of the system composed of 5 wt% of polyoxyethylene (8.6) nonylphenylether, 47.5 wt% of water, and 47.5 wt% of cyclohexane, as a function of temperature. W phase in Fig. 10 below 55 °C is an oil-solubilized aqueous phase which dissolves large amounts of oil and surfactant

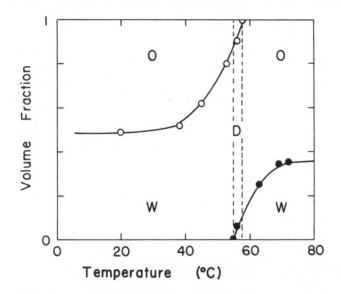

Fig. 10. The effect of temperature on the vol-ume fractions of water, oil, and surfactant phases. The system is composed of 5 wt% of polyoxyethylene (8.6) nonylphenylether, 47.5 wt% of water, and 47.5 wt% of cyclohexane. Reproduced from Saito and Shinoda, Ref. (47), courtesy of Academic Press, Inc.

close to the cloud point (55 °C), and changes continuously to the surfactant phase D. The water phase separated from D phase above 55 °C (cloud point) contains very small amounts of oil and surfactant ($\sim$0.01 wt%). On the other hand, almost pure oil phase O disappears above 58 °C because oil dissolves into D phase. In other words, D phase changes gradually to the oil phase. Such a change can be anticipated from Figs. 2 and 4.

Thus, although the oil-swollen micellar solution (I_w), the surfactant phase in realm III, and the water-swollen micellar solution (I_O) in Fig. 2 are different phases on the phase diagram, the transitions in composition and structure of these phases are quite continuous. Realm I_w is a hydrophilic micel-lar solution in which the surfactant phase is deemed to be a

a lamellar micellar solution of sandwich-like structure whereas realm I is an oleophilic micellar solution in Figs. 2 and
4. The curvature of the surfactant monolayer against oil (or
water) seems to change continuously with temperature, since
the very small interfacial tension between surfactant phase
and water phase (smaller than 10^{-1} dyne/cm) increases with
temperature. On the other hand, the tension between the surfactant phase and oil phase (which is also very small--smaller
than 10^{-1} dyne/cm) decreases with temperature (47). This
finding suggests that the adsorbed surfactant monolayer at the
oil-water interface has a tendency to be concave towards oil
at lower temperatures, is flat at medium temperatures, and
tends to be convex towards oil at higher temperatures. This
results in an o/w-type emulsion at lower temperatures and a
w/o-type at higher temperatures. Although there is no solubilization in the temperature range of realm III because of the
separation of the surfactant phase, the solubility of oil in
the surfactant + water phases, or that of water in the surfactant + oil phases, is high, and detergent action may well
occur at this temperature.

By varying amphiphiles, additives, hydrocarbons, and compositions, Winsor (19,48) was able to define the limits of
completely solubilized systems and the nature of the equilibria of solubilized phases with other phases. He defined equilibria as Type I (solubilized phase in equilibrium with dilute
hydrocarbon), Type II (solubilized phase in equilibrium with
dilute aqueous phase), Type III (solubilized phase in equilibrium with dilute hydrocarbon and dilute aqueous phases), and
Type IV (solubilized phase only). Furthermore, Type IV systems were subdivided into various isotropic sol and birefringent gel phases. Gradual changes in composition led to the
conversion of one system into another. Realms I_w and I_o in
the study correspond to Type IV, realm III to Type III, realm
II_{w-o} to the Type I, and realm II_{o-w} to Type II of Winsor's
classification. The surfactant phase and I_w and I_o phases
adjacent to realm III seem to correspond to Schulman's so-
called microemulsions.

E. The Importance of Distinguishing Various Emulsion Types

There are many types of emulsions besides O/W or W/O.
These are W/D, D/W, D/O, O/D, (W + O)/D, (D + O)/W, (D + W)/O,
O/(D + W), etc., where D denotes surfactant phase. Types of
dispersions comparable to the one in Fig. 2 will be described.
Excess water separates from a nonaqueous micellar solution at
high temperatures. The dispersion type of this two-phase
solution is a W/O type. Over a wide volume fraction range,
the dispersion is a W/O type. In the region where the volume
fraction of oil is smaller than 0.2, the concentration of the

nonionic surfactant is fairly high and the nonaqueous phase
may be considered as a surfactant phase in which hydrocarbon
is dissolved. The solution is viscous at this volume fraction
(about 0.2). If the volume fraction of oil is further decreased, the water phase (which occupies a very large volume
fraction) finally becomes a continuous phase, i.e., phase inversion does occur from a W/D(O) to a D(O)/W type, as shown in
Fig. 11.

At an intermediate temperature three phase coexist, so
that water, oil, and surfactant phases are more clearly distinguished in realm III. The type of dispersion is the
(W + O)/D type; the oil phase disappears owing to the decrease
in the volume fraction of hydrocarbon, so that the type of
dispersion is either a W/D or a D/W type above or below the
phase inversion temperature, in realm II_{d-w}, as shown in Fig.
11.

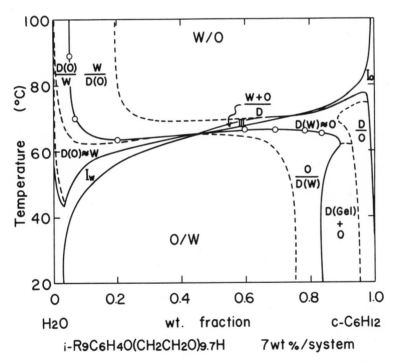

Fig. 11. The effect of temperature and composition on
the dispersion types of the system composed of water,
cyclohexane, and polyoxyethylene (9.7) nonylphenylether
(7 wt%/system). Reproduced from Shinoda and Saito,
Ref. (37), courtesy of Academic Press, Inc.

The change from surfactant phase to oil phase proceeds gradually as the temperature rises as shown by the upper dotted curve. A region exists between the phase inversion temperature and the cloud point curve in which water and surfactant phases are both continuous (W ≈ D).

On the other hand, the water phase in realm III disappears owing to the decrease of the volume fraction of water, so that the type of dispersion is either a D/O or an O/D type above or below the phase inversion temperature. The change from surfactant phase to water phase proceeds gradually as the temperature drops as shown by the lower dotted line. The dispersion in the II_{d-o} realm is not always a D/O type above the phase inversion temperature, but both phases are continuous in one region. The type is clearly a D/O type in the region where the volume of oil phase is large, as shown in realm D/O in Fig. 11.

At low temperatures, the surfactant dissolves in water and some hydrocarbon is solubilized in the aqueous micellar solution. Beyond the solubilization limit excess hydrocarbon disperses as an O/W emulsion. Because of the change of composition from the left-hand side to the right-hand side, the concentration of surfactant in the water phase increases because the amount of surfactant in the system is fixed. Finally, a stiff gel (surfactant phase encompassing water and oil) and oil phases coexist as shown in Fig. 11 in realm D(gel) + O. If the volume fraction of water decreases further, the stiff gel becomes a sol, and a two-phase solution consisting of oil and surfactant phases is obtained.

It may be concluded that the oil phase is the continuous medium at high temperatures, the surfactant phase is the continuous medium at intermediate temperatures close to the phase inversion temperature, and the water phase is the continuous medium at low temperatures in systems composed of water, oil, and surfactant. If we consider that in two phase solutions the surfactant phase in equilibrium with the water phase is an oil phase, and that the surfactant phase in equilibrium with the oil phase is a water phase, the phase inversion temperature vs. volume fraction curve agrees with the normal one (42). In the case of monodisperse nonionic surfactants in which there is no distribution of oxyethylene chain length, the optimum temperature for the solubilization of oil (water) in aqueous (hydrocarbon) solutions coincides with the phase inversion temperature in emulsions.

F. Why w/o Type Microemulsions Are Easier to Prepare than o/w Types

A micelle consists of scores or hundreds of surfactant molecules. The aggregation number of a mixed micelle may be particularly large when a suitable amount of cosurfactant is added to the system. The number of micelles increases with concentration above the critical micelle concentration. Thermodynamic properties such as partial molal Gibbs free energy, enthalpy, and entropy remain nearly constant as a function of concentration. Thus, a micelle may be treated as a pseudophase (5,49) or according to the thermodynamics of micro systems (50). Hence, the distribution of added substances such as hydrocarbons or paraffin chain alcohols in micellar solutions may be considered as partitioning between the micelles and the bulk solution. Thus, solubilization is similar to the solution of solubilizate in micelles.

The basic solubility equation (51) of a perfect solution is

$$\ln a_2 = \ln x_2 + \frac{v_2 \phi_1^2 B'}{RT} \qquad \ldots \ (7)$$

where a_2 is the relative activity of solute in equilibrium with the solute molecule in the micelle as well as in the bulk solution, x_2 the solubility expressed in mole fraction unit, v_2 the molal volume of solute, ϕ_1 the volume fraction of solvent (component I), and B' the energy of mixing per unit volume at infinite dilution. Although Eq. (7) is derived for a perfect solution, the principle can be applied equally well to most solutions.

The relative activity of water is always close to 1 because excess water is considered as pure water (i.e., the vapor pressure of excess water is close to that of pure water). Hence, the amount of water solubilized is roughly proportional to the number and size of micelles and depends on the size and types of hydrophilic groups, but the molecular volume of water (solubilizate), which is very small compared to that of an oil molecule, is fixed. It is evident from Eq. (7) that the molecular volume of solute v_2 is an important factor in controlling solubility.

The amount of oil solubilized is also roughly proportional to the number and size of micelles and not very dependent on the types of lipophilic groups, but is dependent on the molecular volume of the oils. Thus, v_2 in Eq. (7) is small and fixed in the case of w/o type microemulsions so that the solubilization is large and relatively independent of the types of surfactant. On the other hand, in the case of o/w type microemulsions, v_2 is large and varies, so that solubilization is small for oils the molecular sizes of which are large. This is the basic reason why w/o type microemulsions are easier to prepare than o/w types.

Solubilization is, however, a function of the size of micelle, types of solubilizer, the size of solubilizer, the HLB-temperature (40) and HLB-value of solubilizer, the size of solubilizate, temperature, CMC (particularly in nonaqueous solution, because the CMC is not always very small), etc. Factors which increase solubilization, and produce microemulsions with less solubilizer are described in the next section, IV.

IV.   FORMULATING MICROEMULSIONS WITH LESS SOLUBILIZER

In accordance with the foregoing, practical knowledge of microemulsions seems to be able to explain the factors needed to increase solubilization using a lesser amount of solubilizer, to select the optimum solubilizer among many surfactants and to design new types of surfactants as well as cosurfactants which are efficient and applicable under appropriate conditions.

A.   Optimum Temperature for a Given Nonionic Surfactant

It is evident from Figs. 2 and 4 that the solubilization of water (or oil) in nonaqueous (or aqueous) solution of nonionic surfactant exhibits a maximum at the optimum temperature. Thus, the optimum nonionic surfactant, the PIT of which is close to a given temperature, exhibits large solubilizing power which means a microemulsion with less solubilizer. As the optimum temperature for solubilization and the phase inversion temperature change with the hydrophilic chain length of nonionic surfactant (40), an optimum PIT, i.e., HLB-temperature of a surfactant is an important property. For a specific nonionic solubilizer, in other words, one in which the PIT (HLB-temperature) is fixed, that is the optimum temperature for large solubilization.

B.   Optimum Ratio of Surfactants

In order to increase solubilization, the HLB (or PIT) of a surfactant mixture has to be matched to the given oils. Alcohols are often added to solutions of ionic surfactants. Triangular phase diagrams of water, p-xylene, octylamine and octylammonium chloride demonstrate the sensitivity of maximal solubilization to the ratio of the HLB's of the surfactant mixtures (41). An alternative representation (Fig. 7) even more clearly depicts the importance of an optimum HLB ratio.

Sorbitan monoester (18) or $ROCH_2CH_2OH$ (52), which is less oil soluble and a better balanced cosurfactant is more effective, as shown in Fig. 8.

The HLB of nonionic surfactants changes with temperature. Solubilization is large when the PIT (HLB-temperature) of the selected nonionic surfactant is close to a given temperature as shown in Fig. 4. Thus, the optimum HLB ratio in mixtures of nonionic surfactants is an important factor in increasing solubilization, i.e., in preparing microemulsions as shown in Figs. 5 and 6.

C. The Closer the PIT of Two Surfactants, the Wider the Solubilization Range

It became clear from recent studies (18) that the solubilization of oil (or water) is larger, when the nonionic surfactant is monodisperse than when the distribution of hydrophilic chain lengths is broad (commercial material) as shown in Figs. 12 and 13, and when the difference in the PIT's is large. The combination of potassium oleate and alcohol is a mixture of strongly hydrophilic and strongly lipophilic surfactants. The combination of dimethylethanol ammonium (AMP) oleate and sorbitan monododecanoate in which the HLB's of the two surfactants approach each other, showed enhanced solubilization, i.e., roughly twice as high as the former combination as shown in Figs. 1 and 9. A similar relationship holds in the case of aqueous micellar solutions as shown in Fig. 8. Thus, the rule holds equally well for a mixture of an ionic surfactant and a nonionic surfactant.

D. The Larger the Size of the Solubilizer, the Greater the Solubilizing Power

If the size of the hydrophile and lipophile groups of the solubilizer increases, the CMC will decrease, the aggregation number will increase and the solubilizing power will be enhanced. This reasoning is confirmed by changing the sizes of hydrophile and lipophile groups while keeping the PIT (HLB-temperature) of the solubilizer constant. The phase diagram of water and cyclohexane containing 5 wt%/system of $C_{12}H_{25}C_6H_4O-(CH_2CH_2O)_{9.7}H$ is observed as a function of temperature as shown in Fig. 14.

A comparison of Fig. 14 and Fig. 4 demonstrates the extension of realm I_w and I_o with $C_9H_{19}C_6H_4O(CH_2CH_2O)_{8.6}H$ solution in Fig. 4. Experience showed that about 3 wt%/system of $C_{12}H_{25}C_6H_4O(CH_2CH_2O)_{9.7}H$ seemed sufficient to yield a similar microemulsion realm as in the 5 wt%/system of the nonylphenyl compound of Fig. 4.

In order to increase the amount of solubilization as well

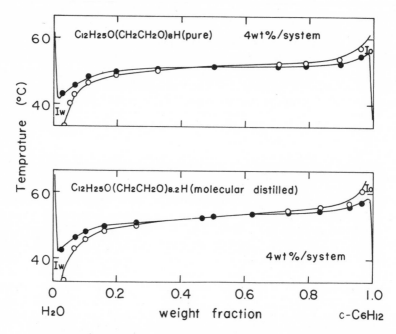

Fig. 12. The phase diagram of water + cyclohexane containing 4 wt%/system of pure $C_{12}H_{25}O(CH_2CH_2O)_8H$ and molecularly distilled $C_{12}H_{25}O(CH_2CH_2O)_{8.2}H$ as a function of temperature. Reproduced from Shinoda and Kunieda, Ref. (18), courtesy of Academic Press, Inc.

as the size of micelle, ionic (53,54) or nonionic (18) surfactants whose hydrocarbon chain lengths are long, should be used. This conclusion agrees with the findings by Gerbacia and Rosano (55), i.e., the longer chain sodium alkyl sulfates require a far smaller amount of alcohol to be added to Schulman's recipes in which the long chain (C_{16} - C_{18}) ionic surfactants were always employed (11,12).

E. Types of Hydrophilic Groups on Surfactants

Since the polyoxyethylene chain is not strongly lipophobic the CMC in the oil phase is not small. Substitution of a polyoxyethylene compound with a suitable mixture of sucrose monoester and sorbitan monoester in the oil phase will decrease the CMC and increase the solubilization of water because both sorbitan monoester and sucrose monoester possess efficient hydrophilic groups of different sizes.

The effect of the substitution of $R_9C_6H_4O(CH_2CH_2O)_{7.4}H$ by

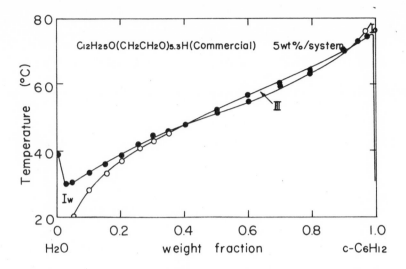

Fig. 13. The phase diagram of water + cyclohexane containing 5 wt%/system of commercial $C_{12}H_{25}O-$ $(CH_2CH_2O)_{5.3}H$ as a function of temperature. The solubilized regions are diminished compared with pure materials. Reproduced from Shinoda and Kunieda, Ref. (18), courtesy of Academic Press, Inc.

a mixture of sucrose tetradecanoate and Span 20 (sorbitan monododecanoate) on the solubilization of water in cyclohexane was studied and the results were plotted in Fig. 15. The solubilization of water increased from 23 to 55 g/100g of solution containing 5 wt% of solubilizer by the substitution (39).

F. Stability to Temperature Change

Ionic surfactants are usually strongly hydrophilic. It is difficult to find a single ionic surfactant whose HLB for a given oil is optimum. Hence, ionic surfactants need a lipophilic cosurfactant to increase solubilization. However, nonionic surfactants change their PIT's (HLB) gradually with their oxyethylene chain length, so that a single optimum nonionic surfactant whose PIT is close to a given temperature can exhibit large solubilizing power. It is evident from the phase diagrams that a nonionic surfactant is a good solubilizer at an optimum temperature but only for a limited temperature range. On the other hand, ionic surfactants are stable to temperature change but need higher concentrations. A mixture of nonionic surfactant and ionic surfactant which is not strongly hydrophilic seems to be ideal (18).

The phase diagram of H_2O + c-C_6H_{12} containing 4.75 wt% of $C_9H_{19}C_6H_4O(CH_2CH_2O)_{7.4}H$ and 0.25 wt% of calcium dodecyl-benzene sulfonate was studied and plotted in Fig. 16. The op-timum temperature for the solubilization of cyclohexane in aqueous solution of $C_9H_{19}C_6H_4O(CH_2CH_2O)_{7.4}H$ is 45°C and $C_{12}H_{25}C_6H_4SO_3\frac{1}{2}Ca$ is an oil-soluble surfactant which contains 30 wt% of methanol.

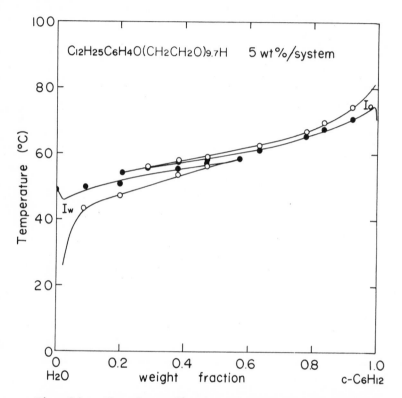

Fig. 14. The phase diagram of water + cyclohexane containing 5 wt%/system of $C_{12}H_{25}C_6H_4O(CH_2CH_2O)_{9.7}H$ as a function of temperature. Reproduced from Shinoda and Kunieda, Ref. (18), courtesy of Academic Press, Inc.

The PIT of this system is close to that of $C_9H_{19}C_6H_4O-(CH_2CH_2O)_{8.6}H$ in Fig. 6, but the solubilization realm is in-creased and the three-phase region is reduced in size by the substitution of nonionics with a small amount of $C_{12}H_{25}C_6H_4-SO_3\frac{1}{2}Ca$. Comparing Fig. 16 and Fig. 17, it is clear that I_w region becomes more stable to temperature change by adding an ionic surfactant. The solution in the turbid realm in Fig.

16 may be called a real microemulsion because the droplets are very small and the solution looks like an emulsion optically, and it separates into two phases upon centrifugation.

The advantage of blending (mixing of surfactants) is that a nonionic surfactant is the main solubilizer and the small amount of oil-soluble ionic surfactant is added to adjust the PIT (or HLB) of the mixture as well as to increase the stability and solubilization. Consequently, the appropriate combination of a balanced ionic surfactant and nonionic surfactant favors effective solubilization and temperature stability.

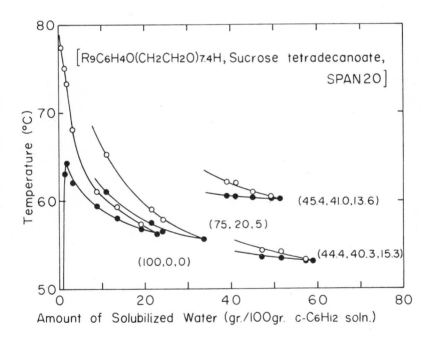

Fig. 15. The effect of the substitution of $C_9H_{19}C_6H_4O-(CH_2CH_2O)_{7.4}H$ with a mixture of sucrose tetradecanoate and sorbitan dodecanoate on the solubilization of water in cyclohexane containing 5 wt%/system of solubilizer. Numbers in parentheses indicate weight percent of respective solubilizer. Reproduced from Shinoda and Kunieda, Ref. (18), courtesy of Academic Press, Inc.

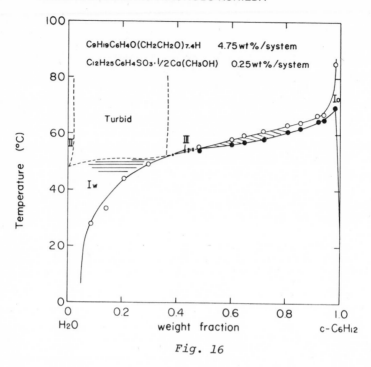

Fig. 16

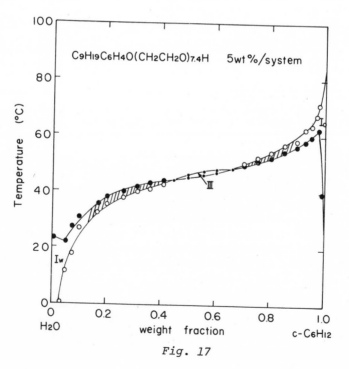

Fig. 17

REFERENCES

1. Lord Rayleigh, *Phil. Mag. 48*, 375 (1899).
2. Einstein, A., *Ann. Physik 33*, 1275 (1910).
3. Smoluchowski, M., *Ann. Physik 25*, 205 (1908).
4. Zimm, B. H., *J. Phys. Coll. Chem. 54*, 1306 (1950).
5. Shinoda, K., "Colloidal Surfactants," Chapter 1, Academic Press, New York, 1963.
6. Shinoda, K., *J. Colloid Interface Sci. 34*, 278 (1970).
7. Lord Rayleigh, *Phil. Mag. 41*, 107, 274, 447 (1871).
8. Schulman, J. H., Stockenius, W., and Prince, L. M., *J. Phys. Chem. 63*, 1677 (1959).
9. Stockenius, W., Schulman, J. H., and Prince, L. M., *Kolloid-Z. 169*, 170 (1960).
10. Prince, L. M., 48th Natl. Colloid Symposium, Amer. Chem. Soc., Preprints, p. 181, June, 1974.
11. Schulman, J. H., and Riley, D. P., *J. Colloid Interface Sci. 3*, 383 (1948).
12. Schulman, J. H., and Friend, J. A., *J. Colloid Interface Sci. 4*, 497 (1949).
13. Cooke, C. E., and Schulman, J. H., *In* "Surface Chemistry," p. 231 (P. Ekwall, K. Groth, and V. Runnstrom-Reio, Eds.), Munksgaad, Copenhagen.
14. Bowcott, J. E., and Schulman, J. H., *Z. Elektrochem. 59*, 283 (1955).
15. Zlochower, J. and Schulman, J. H., *J. Colloid Interface Sci. 24*, 115 (1967).
16. Sherman, P., "Emulsion Science," p. 205, Academic Press, London and New York, 1968.
17. Gillberg, G., Lehtinen, H., and Friberg, S., *J. Colloid Interface Sci. 33*, 40 (1970).
18. Shinoda, K., and Kunieda, H., *J. Colloid Interface Sci. 42*, 381 (1973).
19. Winsor, P. A., *Trans. Faraday Soc. 44*, 376 (1948).
20. Winsor, P. A., *Trans. Faraday Soc. 46*, 762 (1950).
21. Winsor, P. A., *J. Colloid Sci. 10*, 88 (1955).

Fig. 16. The phase diagram of water + cyclohexane containing 4.75 wt% of $C_9H_{19}C_6H_4O(CH_2CH_2O)_{7.4}H$ and 0.25 wt% of $C_{12}H_{25}C_6H_4SO_3{1/2}Ca$ (30% methanol) as a function of temperature. Reproduced from Shinoda and Kunieda, Ref. (18), courtesy of Academic Press, Inc.

Fig. 17. The effect of temperature on the phase diagram of water + cyclohexane containing 5 wt%/system of $C_9H_{19}C_6H_4O(CH_2CH_2O)_{7.4}H$. Reproduced from Kunieda and Shinoda, Ref. (56), courtesy of Chemical Society of Japan.

22. Palit, S. R., Moghe, V. A., and Biswas, B., *Trans. Faraday Soc. 55*, 463 (1959).
23. Shinoda, K., Ed., "Solvent Properties of Surfactant and Solutions," p. 4, Marcel Dekker, New York, 1967.
24. Adamson, A. W., *J. Colloid Interface Sci. 29*, 261 (1969).
25. Tosch, W. C., Jones, S. C., and Adamson, A. W., *J. Colloid Interface Sci. 31*, 297 (1969).
26. Ekwall, P., Mandell, L., and Fontell, K., *J. Colloid Interface Sci. 33*, 215 (1970).
27. Corkill, J. M., Goodman, J. F., and Harrold, S. P., *Trans. Faraday Soc. 60*, 202 (1964).
28. Clunie, J. S., Goodman, J. F., and Symons, P. C., *Trans. Faraday Soc. 65*, 287 (1969).
29. Nakagawa, T., and Shinoda, K., *In* "Colloidal Surfactants," Chapter 2, pp. 121-126, Academic Press, New York, 1963.
30. Kuriyama, K., *Kolloid-Z. 180*, 55 (1962).
31. Becher, P. and Arai, H., *J. Colloid Interface Sci. 27*, 634 (1968).
32. Nord, F. F., Bier, M., and Timasheff, S. N., *J. Am. Chem. Soc. 73*, 289 (1951).
33. Sakurada, I., Sakaguchi, Y., and Ito, Y., *Kobunshi Kagaku (in Japanese) 14*, 41 (1957).
34. Taniguchi, Y., Suzuki, K., and Enomoto, T., *J. Colloid Interface Sci. 46*, 511 (1974).
35. Kunieda, H. and Shinoda, K., *Adv. Chem. Series 9*, 278 (1975).
36. Strauss, U. P., and Williams, B., *J. Phys. Chem. 65*, 1390 (1961).
37. Shinoda, K., and Saito, H., *J. Colloid Interface Sci. 26*, 70 (1967).
38. Saito, H., and Shinoda, K., *J. Colloid Interface Sci. 24*, 10 (1967).
39. Shinoda, K., and Ogawa, T., *J. Colloid Interface Sci. 24*, 56 (1968).
40. Shinoda, K., *J. Colloid Interface Sci. 24*, 4 (1967).
41. Friberg, S., and Lapczynska, I., *Progr. Colloid & Polymer Sci. 56*, 16 (1975).
42. Shinoda, K., and Arai, H., *J. Colloid Interface Sci. 25*, 429 (1967).
43. Ahmad, S. I., Shinoda, K., and Friberg, S., *J. Colloid Interface Sci. 47*, 32 (1974).
44. Shinoda, K., and Friberg, S., *Adv. Colloid Interface Sci. 4*, 281 (1975).
45. Friberg, S., *Kolloid-Z. 244*, 333 (1971).
46. Schulman, J. H., and Montagne, J. B., *Ann. N. Y. Acad. Sci. 92*, 366 (1961).
47. Saito, H., and Shinoda, K., *J. Colloid Interface Sci. 32*, 647 (1970).
48. Winsor, P. A., "Solvent Properties of Amphiphilic Compounds," Butterworths, London, 1954.

49. Shinoda, K., and Hutchinson, E., *J. Phys. Chem. 66,* 577 (1962).
50. Hall, D. G., and Pethica, B. A., *In* "Nonionic Surfactants" (M. J. Schick, Ed.), pp. 516-557, Marcel Dekker, New York, 1967.
51. Hildebrand, J. H., Prausnitz, J. M., and Scott, R. L., "Regular and Related Solutions," Van Nostrand Reinhold Co., New York, 1970.
52. Shinoda, K., To be published by *J. Colloid Interface Sci.*
53. Stearns, R. S., Oppenheimer, H., Simon, E., and Harkins, W. D., *J. Chem. Phys. 15,* 496 (1947).
54. Klevens, H. B., *Chem. Rev. 47,* 1 (1950).
55. Gerbacia, W., and Rosano, H. L., *J. Colloid Interface Sci. 44,* 242 (1973).
56. Kunieda, H., and Shinoda, K., *Chem. Soc. Japan (in Japanese),* 2001 (1972).

The Mixed Film Theory

LEON M. PRINCE

Consulting Surface Chemist
7 Plymouth Road
Westfield, New Jersey   07090

I. INTRODUCTION

In the case of microemulsions, as in many other developments, the art preceded the science. Hoar and Schulman (1) in the first scientific paper on the subject acknowledged this by referring to the "soluble oils" of commerce--the cutting oils of the machine tool industry. Schulman in England was not aware of the development in the United States of the Carnauba wax and other emulsifiable waxes for floor polishes, nor of the flavor oil, chlordane, alkyd and dry cleaning emulsions, all of which were microemulsions in the true sense but not recognized as such. In 1958 when he came to Columbia University to accept the chair as Stanley Thompson Professor of Chemical Metallurgy in the School of Mines, the writer met him and told him of these developments with which the writer was familiar or had worked with for twenty years. These products interested Schulman greatly, and with Dr. Walter Stoeckenius of the Rockefeller Institute he was responsible for three papers in which, among other things, the idea of staining the alkyd emulsions of commerce with Osmium tetroxide was conceived and reduced to practice (2,3,4). This led to the micrograms shown in the frontispiece and prompted Schulman to coin the term "micro emulsion" to describe these systems. In the second paper, at the insistence of the writer and as a courtesy to him, the term was contracted to one word. The two terms were used interchangeably for a few years but with time the word "microemulsion" persisted.

By 1955 Schulman had begun to use only w/o microemulsion systems in his experimental work because they were easier to find and make, and were susceptible to treatment which enabled him to establish the ratio of alcohol (amphiphile) to soap at the interface. During our collaboration it was gradually recognized that Schulman thought in terms of w/o systems and the writer in terms of o/w systems. Certain adjustments in theoretical concepts were made to accommodate these different viewpoints since it was felt that both the w/o and o/w systems were the same type of colloidal dispersion, although probably formed by different kinds of interactions in the interface. The fact that some of Schulman's w/o microemulsions, notably Bowcott and Schulman's benzene-in-water systems (5), failed to invert through a viscoelastic gel stage to o/w microemulsions presented something of a problem but was not considered a serious one at that time.

Schulman's view of these systems stemmed directly from his work with mixed films exemplified by those of the small droplet sized emulsions formed in the famous papers by Schulman and Cockbain (6). He saw the o/w system as an alternative to association as curd fibres and the w/o system as an analogue to Lawrence's hydrophilic swollen, soap, ionic micelle

containing enclosed oil (1). Subsequently (2), he theorized
that soap or detergent micelles possessed an array of molecules too ordered to expand in the presence of a nonpolar compound beyond certain small limits (10%). This indicated that
the monolayer of the bimolecular leaflet of the (lamellar)
micelle aggregate was not truly in a liquid state and thus
able through surface tension forces to take on a curvature.
Such curvature could be achieved in a mixed film which was
liquefied due to the presence of alcohol, by performing a
minimum of work on the system, in accordance with Bancroft's
theory (7,8). But when nonpolar oil molecules could also
penetrate the film, the curvature occurred spontaneously.
This strengthened Schulman's belief that these small droplet
sized dispersions were emulsions. Nevertheless, he appreciated that such systems, without hydrocarbon, could form types
of liquid crystals or myelinic structures (9).

In 1955 with Bowcott (5), Schulman postulated that the
interface was a third phase or interphase, implying that such
a monolayer could be a duplex film, i.e., one having different
properties on the water side than the oil side. Such a specialized liquid, two-dimensional region bounded by water on
one side and oil on the other was a new concept to which
Schulman and his co-workers soon addressed themselves with
rewarding results. These were based on the assumption that
the spontaneous formation of a microemulsion was due to the
fact that the interactions in the interphase among soap, alcohol and oil phase molecules reduced the original oil/water
interfacial tension to zero.

In this perspective, one could explain the formation of a
microemulsion in terms of the molecular interactions in the
interphase. At the concentration of alcohol and soap (including penetration by oil phase molecules) required for zero
interfacial tension, the ratio of the volume of interfacial
species to the volume of dispersed phase is high enough to
occupy the interfacial volume needed for droplets having diameters less than 1/4λ of light. However, zero interfacial
tension alone does not ensure that a microemulsion will form
in these oil, water, and surfactant systems since cylindrical
and lamellar micelles also exist in the equilibrium state.
What differentiates an emulsion from these liquid crystalline
phases is the kind of molecular interactions in the liquid
interphase that produce an initial, transient tension or pressure gradient across the flat interphase, i.e., a duplex film,
causing it to enclose one bulk phase in the other in the form
of spheres. When mechanical work is required to effect such
curvature, Bancroft called the system an emulsion; when the
curvature occurred spontaneously, Schulman called the system a
microemulsion (10).

Schulman's approach to this work was experimentally circum-
spect. In the course of his twenty-five year (1943-1967)

investigation he left no appropriate tool unused in his quest
to identify these colloidal dispersions. The critical review
which follows testifies to his diligence and imagination.

Before approaching such a review, in all fairness a few
philosophical observations are in order. The development of a
theory as complex as the one needed to explain microemulsions
or micellar solutions is achieved in a step-by-step fashion.
Many hypotheses are made, most of which fall by the wayside as
experimental evidence is obtained to indicate that they are
untenable. Even then, the conclusions may be only partially
correct. Time alone tells. It was Jack Schulman's credo that
few men are privileged to postulate the whole truth in their
lifetime. In this context, the writer has arbitrarily omitted
some of the hypotheses which, from the vantage point of 1976,
appear to have been in error, and has attempted to indicate
the step-by-step route by which the present state of the theory has been arrived at. In this perspective Schulman's path
was more straight than zig-zag.

II. THE EARLY YEARS

During the first decade of his study of these transparent
or translucent dispersions of oil, water, and surfactant,
Schulman debated in his own mind whether these small aggregate
size dispersions were emulsions or micelles. In the first
paper (1), the w/o dispersions were called oleopathic hydromicelles. Later (11) he called them oleophilic hydromicelles
and the o/w dispersions, hydrophilic oleomicelles. He entitled
his papers "Studies in Water and Oil Dispersions" or "Penetration and Complex-formation in Monolayers." It was not until
1955 with Bowcott that a paper was entitled "Emulsions."

Transparent oil, water, and surfactant systems were first
identified by Hoar and Schulman as a special kind of colloidal
dispersion. The opening paragraph of their letter to the editor of *Nature* bears quoting for its directness: "It is well
known that oil-alkali metal soap (or cationic soap, such as
cetyl trimethyl ammonium bromide)-water systems of certain concentrations exist as transparent, electrically non-conducting
dispersions, in which the oil is the continuous phase. Dilution of these systems with excess water inverts them to oil-in-
water emulsions which are milky for low soap/oil ratios and
transparent for sufficiently high soap/oil ratios. The transparent oil-continuous systems are familiar as "soluble-oil"
and similar concentrates; the essential conditions for their
formation are: (1) high soap/water ratio, (2) the presence of
an alcohol, fatty acid, amine or other non-ionized amphipathic
substance in mol-fraction approximately equal to that of the
soap." They further indicated that the disperse phase consisted of submicroscopic micelles surrounded by a mixed

monolayer of soap and amphipath. Their calculations of the size of w/o systems displaying a slight Tyndall effect was in the 120 A to 200 A range and the vertical depth of the soap monolayer, 18 A. It was observed that addition of water to the transparent oil-continuous systems turned them into opaque and viscous systems which finally inverted to stable oil-in-water dispersions. They attributed the spontaneous emulsification to very low interfacial tension.

In order to study the role that alcohol molecules played in enabling the aqueous soap systems to absorb large quantities of oil or water and how they controlled oil or water phase continuity, Schulman and McRoberts (12) added a group of aliphatic and cyclic alcohols to benzene, paraffin and mineral oil. It was found that the number and arrangement of the carbon atoms, both in the alcohol and oil, determined whether the system was w/o or o/w. In this work the mixtures were titrated to transparency with the alcohols. Low angle X-ray scattering analyses were begun on these systems. With these data and on the basis of monolayer penetration experiments it was now deduced that the area per pair of soap-alcohol molecules in the mixed film was 60 A and that the thickness of the interfacial film was 24 A. The area per soap molecule in the film was estimated at 30 to 40 A, depending upon its degree of dissociation in contact with water. Perhaps the greatest contribution of this paper was the discussion of order and disorder in the interfacial monolayer and the bearing of this upon the kind of emulsion that was formed. The penetration of the original soap film was considered in terms of the configuration and length of the alcohol molecules and the strength of their adhesion to the soap molecules, it being essential that the state of the mixed film be liquid in order for dispersed droplets to form.

In the next two papers (13,14) the size of the dispersed aggregates was studied by means of low angle X-ray measurements. Droplet sizes in the range of 100 to 600 A were found. It was noted that water droplets gave stronger, less diffuse X-ray scattering than oil droplets. The 600 A diameter droplets displayed strong Tyndall scattering. In conclusion, the hope was expressed that by these means the gap between swollen micelles and emulsions could be bridged and that the X-ray and microscopic data could be correlated with the structure of these colloidal dispersions.

Two statements in this period have important implications to the future work on micellar solutions and microemulsions, respectively. In reference (13) an alcohol was equated to a polar oil, suggesting the possibility of ternary systems. And in the paper immediately following the Schulman and McRoberts' one in *Nature*, R. C. Pink (15) proposed that, if the assumption that a soap film could be liquefied by penetration with an alcohol, the necessary disorder in the micelle could be produced by temperature alone. He proceeded to demonstrate this with a study of the critical effect of temperature on the absorption of water by solutions of ethanolamine oleate in benzene.

The next tool that Schulman utilized to elucidate the structure of these colloidal dispersions was light scattering. (The apparatus was built by Paul Doty.) In an address before the Society of Cosmetic Chemists on November 9, 1949, the text of which was published a month later in the first volume of the journal of the Society (11), Schulman reviewed the progress made. Much of this was based on the work by Schulman and Friend (16).

It was pointed out that these fluid, isotropic dispersions of droplets of oil dispersed in water and droplets of water dispersed in oil, were the result of a mixed monolayer of discrete molecules adsorbed at the interface between the oil and water phases. Phase continuity and the size of the droplets depended upon the number of carbon atoms and their arrangement in the amphipath. The formation of the dispersion was visualized as commencing with a lamellar micelle in strong soap solution. This micelle could be partly swollen with oil without breakdown of the crystal lattice. This lattice, however, could be further penetrated by small amphipathic molecules like alcohols, causing the mixed film to become liquid and allowing surface tension forces to form spherical droplets of water or oil. On decreasing the soap concentration, the droplet would swell, creating a lesser total surface area and a larger droplet size. Lack of optical streaming birefringence established that the aggregates were spherical.

In view of these properties, it was concluded that the structure of the fluid, isotropic dispersions was different from that of concentrated solutions of soap and of so-called "solubilized oils." These could swell only within certain definite limits (10%), and Schulman believed that the disorder introduced into the interfacial film by the alcohol enabled the shell to enclose a larger volume of dispersed phase. These concepts are well illustrated in Fig. 1.

Viscosity changes observed to accompany the addition of water to the w/o dispersions introduced a new complexity. A transparent birefringent phase had been detected by Winsor (17) in similar systems and caused Winsor to discount the idea of a spherical droplet surrounded by a mixed film of soap and alcohol. Schulman countered with the suggestion that those systems giving anomalous electrical conductivity, streaming birefringence, and high viscosity, i.e., the viscoelastic inversion stages, were composed of lamellar or cylindrically shaped aggregates, in other words, were the liquid crystalline phases that Winsor saw.

In 1951 with Matalon and Cohen (18), he developed evidence for the existence of lamellar, cylindrical, and spherical

micelles in his systems. Solutions of long chain nonionics
(alkanol-ethylene oxide adducts) dissolved in petrol ether
were titrated with small quantities of water and the structure
of the systems were investigated by X-ray, optical, and rheological means. In the absence of water, the systems exhibited
no structure. Upon addition of water, and depending upon the
hydrophilic-lipophilic balance (HLB), as well as the number of
different species of nonionics (mixed emulsifiers), lamellar,

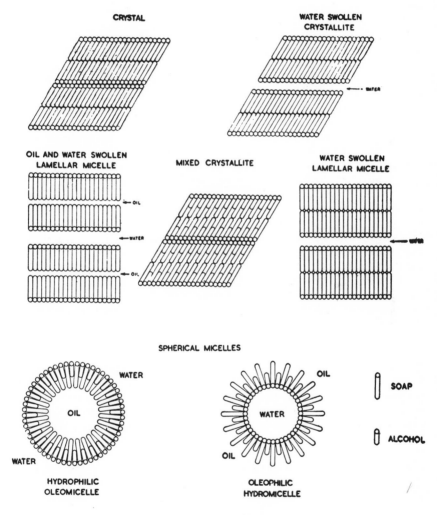

*Fig. 1. The phenomena of solubilization of soap crys-
tal lamellae by nonpolar oils and penetration of the
monolayer lattice by polar oils in the formation of
micelles and emulsions. Reference (11), Fig. 4, courtesy
of the Society of Cosmetic Chemists.*

cylindrical, and spherical aggregates were observed; the several forms being differentiated by means of X-rays. With minimal disorder in the lattice of the aggregates, lamellar bimolecular leaflets appeared in which there was a maximum 1:1 ratio of water to ethylene oxide chain. The structure was related to the interplay between water molecules and the ethylene oxide groups attached to the aliphatic tails and the ratio of the length of the aliphatic group to the number of ethylene oxide groups in the adduct. The strongly birefringent, viscoelastic gels of the lamellae became heterogeneous upon further addition of water. With medium degree of disorder in the lattices, double X-ray band diffraction signalled the presence of an hexagonal array of cylinders in the gel. By luck, in one instance, an isotropic system with a typical diffuse X-ray pattern was observed which was interpreted to indicate the presence of a w/o emulsion, the interfacial species of which were in a considerable state of disarray. It was proposed that as disorder among the nonionic aggregates increased, the lamellae changed into cylinders and then into spheres. These, in turn, were in temperature dependent equilibrium with the lamellae or cylinders. The birefringent systems of lamellar or cylindrical structure were viscoelastic, whereas the isotropic system containing spheres was "very thin fluid."

III.   THE BOWCOTT AND SCHULMAN PAPER

In 1955 in the paper entitled "Emulsions, Control of Droplet Size and Phase Continuity in Transparent Oil-Water Dispersions Stabilized with Soap and Alcohol," Bowcott and Schulman (5) made an important contribution to the theory of microemulsions. Schulman believed this to be his best paper to date on the subject. The tattered condition of the writer's autographed copy of this reprint indicates the number of times he has referred to it, each time with more rewarding insight. Although the results of the ultracentrifuge study were not very enlightening, it appears that for the first time Schulman saw his dispersions in the perspective of emulsions albeit not of the classical type.

He attributed the formation of these emulsions to the molecular interactions taking place in the interface. This mixed monolayer he saw as a third phase or interphase in equilibrium with the oil and water phases. Thus, the alcohol was distributed between the three phases and the whole of the potassium oleate was in the interphase. The last assumption was limited. It was not valid when there was insufficient water present to provide a large enough interfacial area for the whole of the oleate soap to be in the interphase. From previous work, it appeared that 100 A was about the smallest diameter of water droplets which could form microemulsions. This 100 A minimum coincides with that proposed by Prince from other considerations (10).

A liquid condensed film was considered essential to give the kind of flexibility to the interphase that would allow a tension gradient across it to produce curvature. The state of the film was considered to be controlled by the kind of adlineation among the film tenants. Among these factors he now included the size of the cation in anionic systems, and temperature.

A calculation of the ratio of alcohol to soap in the interphase was made in an ingenious way. First, the original w/o emulsions were made by titrating coarse emulsions of hydrocarbon and water stabilized with soap, to transparency with an alcohol. Then additional oil was added and the systems were titrated to transparency again with more alcohol. Repeating this process determined the alcohol volume per hydrocarbon volume to produce a transparent system. Extrapolation of the curve to zero hydrocarbon determined the amount of alcohol in the interface.

The idea of plotting the mole ratio of alcohol:soap vs. the mole ratio of oil:soap in these w/o systems paid other dividends. It was found that to dilute the system and maintain transparency it was necessary to keep a constant benzene to alcohol ratio in the continuous phase. The curves were straight lines so that the intercept with the axis could be established with some certainty. Also the fact that the curve was a straight line was interpreted to confirm that water existed as discrete droplets in these transparent systems but that the interfacial area was not uniquely determined by a fixed mass of soap. In addition, curves were drawn for varying amounts of water, indicating that the ratio of alcohol to soap molecules in the interface increased with the water to soap ratio. This alcohol: soap ratio reached a value of 3.2 before it was impossible to clear the system with hexanol. Systems stabilized with higher alcohols proved to be more restricted as to the amount of water they could solubilize. Beyond decanol, no microemulsion formed. With decanol the ratio of alcohol to soap in the more rigid interface was 1:1. It appears to be significant even at this late date that no micromicroemulsions form when the alcohol:soap ratio is less than 1:1. These experiments left little doubt as to the role that molecular interactions in the film played in controlling droplet size.

The contents of this paper has formed the basis of much discussion by many authors. A number of individual statements in this paper justify Schulman's high appraisal of it. Several of these are listed below.

Previous results of monolayer penetration experiments at the air/water interface (19) made it evident that alcohol

molecules penetrated the adsorbed soap molecules and disordered the regular condensed two dimensional packing in micelles to produce a liquid interphase at the oil/water interface. This enabled Bowcott and Schulman to state that the
interphase had two interfacial tensions, one between the monolayer and the oil and one between the polar heads of the monolayer and the water. The side of the monolayer with the
higher tension would be the inner surface. This was not unlike the duplex film for which Langmuir (20) developed equations and was the same explanation as Bancroft had given for
the formation of two types of macroemulsions.

Schulman recognized that the effect of forming a complex
between a water soluble soap and an oil soluble amphipath reduced the interfacial tension between oil and water to a fraction of a dyne, and it was for this reason that the systems
formed spontaneously. In the Bowcott and Schulman experiments
he noted that the same transparent emulsions were obtained irrespective of the order in which the components were added.
This prompted him to propose that the phases were in equilibrium with each other, inferring that the interfacial tension
was zero!

The measurement of droplet size continued to be a problem.
The most reliable scheme was the calculation of droplet radius
r by means of the formula

$$r = 3V/A$$

where V is the total volume of the dispersed phase and A is
the total interfacial area. This presupposed a knowledge of
the value of the area of the oleate molecule in the interphase
and the assumption that all of it was there. In this case, A
equalled $6X10^{23}$ (the weight of the oleic acid divided by its
molecular weight and multiplied by its area in the film). Now
the area of an oleate molecule at the oil/water interface had
been determined at 30 A^2 and that of an aliphatic alcohol at
$20^2 A$. A 1:1 ratio in the interface should account for an
oleate area of about 50 A^2. Experience, however, led Schulman
to assign a value of 70 A^2 to the area of an oleate molecule
at the oil/water interface. This could be accounted for by an
alcohol:soap ratio of 2:1 or, as proposed later, by penetration of molecules derived from the oil phase. Using this value of 70 A^2 for the area of the oleate molecule, droplet diameters ranging from 140 to 240 A were obtained for the Bowcott and Schulman w/o emulsions. By taking the volume of the
dispersed phase to include the core plus the interphase, it
was not necessary to make any assumptions about chain lengths
nor their orientation other than that the mixed film be in a
completely liquid state.

In the Bowcott and Schulman paper as in the Hoar and

Schulman letter to the editor, a viscoelastic stage was encountered. As the coarse emulsions were titrated with alcohol,
a viscous, birefringent and transparent system occurred. This
suggested again that in this stage the dispersed aggregates
were in the form of long cylindrical micelles or lamellae.

Finally, work with pentanol and butanol, both having solubility in water, also gave straight line curves indicating
that they were w/o emulsion systems. They, however, possessed
abnormally high conductivities until they were well diluted
with benzene. The fact that these systems could be diluted
with the nonpolar oil and yet have high conductivity led Schulman to surmise that the continuous phase, although predominantly benzene, might in fact be a ternary <u>solution</u> containing
water. Here, Schulman was saying that ternary solutions may
exist in these systems but that they occur with different
interfacial species than are responsible for the formation of
microemulsions.

IV. THE OIL/WATER INTERFACE

Until 1958 Schulman considered the interactions among the
molecules at the oil/water interface as closely resembling
those at the air/water interface. As it turned out, there are
important differences that have special significance to the
formation of microemulsions. These differences are the subject of this section.

There was a prelude to this conceptual change that had its
origins in both art and theory. The writer had closely followed Schulman's work in England and had obtained photostatic
copies of references (1), (6), and (12) from the New York Pub-
lic Library. In the United States during the period 1939-47
it was of commercial interest to discover how to form these
stable, opalescent systems so that a substitute could be found
for the increasingly expensive Carnauba wax used in floor polish and carbon paper. It had been noted with more than casual
interest by the writer that both Carnauba wax and Ouricury wax
were unusual among the natural waxes in that they both were
emulsifiable and possessed high hydroxyl values. This seemed
to imply that some component or components of these waxes were
alcohols and were forming complexes with the conventional soap
emulsifiers used in the floor polish emulsions. In late 1941
the writer made a microemulsion of Paraffin wax (M.P. 110°F)
by blending it 3:1 with Myricyl (a C_{31}) alcohol extracted from
Carnauba wax. In this same year, Zisman (21) suggested that an
alcohol monolayer at an oil/water interface could be penetrated
by molecules of the oil phase. Subsequently, a patent was
granted to the writer (22) in which it was shown that mixtures
of Ouricury wax and petroleum hydrocarbons could be microemulsified if the hydrocarbon were blended with it in certain proportions. These proportions were linearly related to the

molecular weight of the hydrocarbon. This led to further experimentation with purer hydrocarbons, natural and synthetic
waxes, and alcohols of various kinds which confirmed a relationship in terms of molecular structure, allowing oil molecules to find their way into the interface. This was how the
stable, opalescent o/w emulsions were formed.

This was the topic of the first meeting between Professor
Schulman and the writer at the Columbia School of Mines in
August of 1958. Schulman at first disapproved of the idea
that oil molecules could penetrate a mixed film of soap and
alcohol but a fortuitous demonstration in the laboratory
changed his mind. A number of long discussions followed which
led to the Schulman, Stoeckenius, and Prince paper (2). It
was Schulman's idea to stain the writer's alkyd emulsions with
osmium tetroxide. Stoeckenius, at the Rockefeller Institute,
obtained remarkable pictures on the very first attempt. Four
of these micrograms and the writer's interpretation of how
Carnauba wax formed a microemulsion were published in reference (4). By that time, the key alcohols in Carnauba wax were
shown by Murray and Schoenfeld in Australia (1955) to be homo-
logous, long chain esters of omega hydroxy acids (4). They
comprised more than 50% of the weight of the wax.

As the writer fed the commercial information to him, Schulman proceeded to put the empirical findings of the past 30
years in the context of surface chemistry. There were a number of different aspects to this.

He noted from Zisman's work that an alcohol layer at the
oil/water interface could be penetrated by long chain mineral
oil molecules to form a 1:1 association and that surface pressure as high as 30 dynes/cm failed to squeeze the nonpolar
hydrocarbons out. At the same time he observed that benzene
could easily be ejected from a monolayer of fatty acids at the
benzene/water interface. Moreover, he knew from the work of
Robbins and LaMer (23) that the solvent hydrocarbons used to
spread monolayers at the air/water interface could remain in
the film but also could at moderate pressures be squeezed out
and not form associations even at low surface pressures. From
this he deduced that if there was a possibility for the oil
molecules to associate with the tenants of the interfacial
film, a microemulsion would form. The association could take
place with either the alcohol or soap tail but the microemulsions were better if the association was with both.

It was now evident that a vapor, rather than a liquid, condensed film was essential to the development of microemulsions
and that this state could be brought about in a number of ways.
The first was the penetration of a mixed film of soap and alcohol (or their equivalents) by hydrocarbons derived from the
oil phase. The second was to use large cations to make the
soap molecules asymmetric and thereby produce disorder in the

mixed film. Finally a microemulsion could be produced with asymmetric soap molecules (without associating alcohol) providing the film were penetrated by oil molecules which associated with the soap species but were sufficiently asymmetric therewith to produce the required disorder. Emphasis was placed on the size of the cation in effecting disorder.

This was of some consequence since in the late thirties, Commercial Solvents had developed 2-amino-2-methyl-1-propanol (AMP), offering it to the wax trade as a replacement for the ethanolamines. Although it was on the expensive side, the writer found it to be the most effective emulsifying cation available; for example, its soaps could microemulsify Candililla wax and other difficultly emulsifiable oils with ease. To explain its performance Schulman proposed that AMP should be considered as 2-amino-2-dimethyl-1-ethanol, a more symmetrical formula. Armed with the further information that in the commercial wax polishes at least twice the mol ratio of cation to anion was required to form stable emulsions, Schulman and Montagne (24) proposed that the AMP molecules formed a structural lattice in the water phase adjacent to the monolayer. Only every other amino group was charged to act as a counterion to the soap anions but all the hydroxyl groups at the other end of the molecule were linked together by hydrogen bonding. This formed a lattice which permitted hydrocarbons to penetrate the monolayer and allow its molecules to readily form a vapor condensed interphase which could then envelop one phase in the form of microdroplets.

There was one further piece of information derived from the art of which Schulman made capital. In addition to twice the mol ratio of cation to anion, wax formulators traditionally used borax in their polishes. This was considered a cheap filler, but the writer early realized that borax, $Na_2B_4O_7,10$ H_2O, decomposed to NaOH and boric acid in hot aqueous solution. Figuring (correctly or not) that boric acid would free some oleic acid from the soap and thereby lower its HLB (25,26), the writer had made a transparent kerosene-in-water emulsion stabilized with AMP-oleate, without the use of a long chain alcohol, by using a small amount of boric acid.* Schulman explained this experiment in much more sophisticated terms. The efficiency of soap as an emulsifier is pH dependent. At pH greater than 10.5 all the soap is in the form of oleate ion and sodium ion. At pH 8.8, the ratio of free fatty acid to oleate ion is 1:1 and at pH 6.8 the ratio of free fatty acid to oleate ion is 2:1 (27). The vital point is that free fatty acid acts very similarly to an alcohol in a monolayer. Schulman called this the formation of amphiphilic agents *in situ*, and it became the basis of the experimental work upon which

---

*It is of interest that this experiment has been performed successfully only when AMP is the cation.

many of the theoretical inferences in this paper were drawn.

Following closely on the heels of the foregoing came the concept of a transient negative interfacial tension as the factor being responsible for the formation of microemulsions (3). During lunch one day the writer suggested that thermodynamic considerations had been omitted from our speculations regarding the formation of microemulsions. Within 15 minutes, Schulman came up with the idea of a transient, negative interfacial tension as producing the energy needed for their spontaneous formation. Essentially, he based this on two bits of information. In the monolayer studies with Goddard (19) two dimensional surface pressures of over 50 dynes/cm had been recorded. In addition, Bowcott (28) had shown experimentally that microemulsions formed at concentrations of surface active agents in excess of those necessary to produce zero interfacial tension against $\log_{10}$ of the mol fraction of hexanol in benzene (29).

The negative interfacial concept was based on the thermodynamic equation that at the oil/water interface

$$\gamma_1 = \gamma_{o/w} - \Pi \qquad (1)$$

where γ_1 is the total interfacial tension, $\gamma_{o/w}$ is the oil/water interfacial tension without the addition of amphipathic agents, and Π is the two dimensional spreading pressure of the amphipathic agents. According to this equation, if as a result of the adsorption of soap and alcohol at the interface and its penetration by oil phase molecules, Π becomes greater than $\gamma_{o/w}$, then energy $-\gamma_1 dA$ (A = surface area) would be available to increase the total interfacial area. This was considered to be the condition for the formation of a microemulsion. When $\gamma_{o/w} > \Pi$ only a macroemulsion could form. In this view the temporary existence of a film pressure greater than $\gamma_{o/w}$ would be the driving force which reduced the droplet size of the fixed volume of dispersed phase until no more energy was available to increase the interfacial area (i.e., to decrease droplet size). Equilibrium would be attained when the negative interfacial tension returned to zero by virtue of the uncrowding of the molecules and the loss of pressure in the interface. When $\gamma_{o/w} > \Pi$, droplet diameters of the order of magnitude of 10,000 Å (1 μm) were usually observed, and the systems which now appeared milky white, achieved equilibrium by separating into two phases. Energy in the form of mechanical work (agitation or homogenization) may temporarily increase the interfacial area but is not capable per se of changing the values of Π or $\gamma_{o/w}$. Schulman summed it up nicely when he said that negative interfacial tension produced by the mixing of the components would, at equilibrium, become zero and dispersion, and not separation, would be the

equilibrium condition. With $\gamma_{o/w}$ for n- paraffin being about
50 dynes/cm and that for benzene, 35 dynes/cm, a Π of greater
than 55 dynes/cm as measured in reference (19) would easily
account for the experimental results obtained.

Unfortunately, this was not the whole truth. It was soon
clear that eq. (1) required modification when Cooke and Schulman (29) determined experimentally that hydrocarbons would be
ejected from mixed monolayers of soap and alcohol at the high
pressures necessary for negative interfacial tensions. These
authors, by using different hydrocarbons instead of different
alcohols or amounts of water as in the Bowcott and Schulman
experiments, found that the distribution of hexyl alcohol
between the bulk phase and the interphase varied with the
hydrocarbon used. The curves were straight lines again, and
the intercept with the vertical axis gave the ratio of alcohol:
soap in the interphase of each oil and water system. This
clue prompted Prince (30) to propose that negative interfacial
tension in mixed films of soap and alcohol is the result not
so much of a high value of Π as of a large depression of $\gamma_{o/w}$.
The new and much lower value of the oil/water interfacial tension, $(\gamma_{o/w})_a$, is dependent upon the amount of alcohol left in
the bulk oil phase after the chemical potential of the alcohol
in each phase has been equalized by partitioning.

The term $(\gamma_{o/w})_a$ introduced a new dimension into behavior
at the oil/water interface. It was soon recognized that it
could asymptotically approach a value of 15 dynes/cm at relatively low concentrations of alcohol in the oil phase. Essentially, the distribution or partitioning of alcohol between
the interphase and the bulk oil phase, changed the composition
of the oil phase and so, its tension with water. Indeed, the
curve of $(\gamma_{o/w})_a$ with concentration of alcohol in the bulk oil
phase could be a steep one with the appropriate molecules, as,
for example, with heptane and cetyl alcohol.

By reason of this, the film pressures needed to reduce the
net interfacial tension to negative values were much lower and
more easily attained. The equation representing this new
state of affairs became

$$\gamma_1 = (\gamma_{o/w})_a - \Pi \qquad (2)$$

As a corollary to this concept, it became apparent that in
any given system, zero interfacial tension may occur only at
an intermediate concentration of alcohol, amphiphile, or cosurfactant (10). This was empirically illustrated in Fig. 1
of reference (30). The explanation is that below this intermediate range, Π may be high but $(\gamma_{o/w})_a$ has not yet been sufficiently depressed to result in a negative tension. Above
the intermediate range, the predominantly alcoholic interphase
may, depending upon the structure of the molecules involved,

either squeeze the oil molecules out of the interphase making $\Pi < (\gamma_{o/w})_a$ or, even with oil molecules still present, become too rigid to develop curvature because of the strong attraction among the heads and tails of the tenants.

Another corollary of the negative tension hypothesis offers an explanation for the stability of microemulsion systems (24, 29,31). This is due to zero surface free energy. When two microdroplets coalesce to form a droplet of larger size, the interfacial tension of the new droplet becomes negative. The large droplet now spontaneously increases its interfacial area to effect zero surface free energy and two droplets of the original size form once again. Brownian Movement no doubt aids and abets this equilibrium. It is this thermodynamic (if not kinetic) equilibrium that keeps the emulsion stable.

The bending of the interphase both in the context of thermomodynamics and molecular interactions could now be represented on pressure area curves (32). Bancroft and Schulman saw the interface as a duplex film with different tensions on either side of it. But, if only the freshly adsorbed, <u>flat</u> film was considered as duplex, then curvature of the film under the stress of the tension gradient, would make the pressures or tensions on both sides of the <u>curved</u> film the same. This mechanism is illustrated graphically in Fig. 2. Here, the film/oil surface and film/water surface of an o/w emulsion are both characterized by their own hypothetical (Π-A) curves (cf. Chapter 3). Curve EF is the actual relationship of a mixed duplex film measured on a Langmuir trough (24), and is the sum of curves AB and CD. Under these circumstances Π_w' and Π_o' are the film pressures of the flat duplex film at the water and oil side, respectively, and Π_w and Π_o are the corresponding pressures at the sides of the curved film. The initial pressure gradient, $\Pi_{\overline{G}}$ across the flat film derives from the relative magnitudes of Π_w' and Π_o'. A reasonable value of the area per fatty acid molecules including the area of the alcohol, but prior to penetration by molecules of the oil phase, could be about 50 A^2, as shown. Values of Π_w' could be 30 dynes/cm and for Π_o', 10 dynes/cm. Under the stress of these pressures plus that due to the penetration of oil molecules, expansion at both sides of the interphase would spontaneously take place. This expansion would continue, to different degrees at each side of the interphase, until these pressures became equal and the total pressure Π in the film dropped to $(\gamma_{o/w})_a$. Since $\Pi = \Pi_o + \Pi_w$, expansion would occur until $\Pi_o = \Pi_w = 1/2(\gamma_{o/w})_a$, a reasonable value for which was deemed to be 7.5 dynes/cm. Graphically what happened was that the pressure at the water side slid down curve AB from 30 dynes/cm to its final value of 7.5 dynes/cm and the initial pressure at the oil side has slid along the curve CD from 10 to 7.5 dynes/cm. By the terms of eq. (2), the driving force for this behavior is the pressure

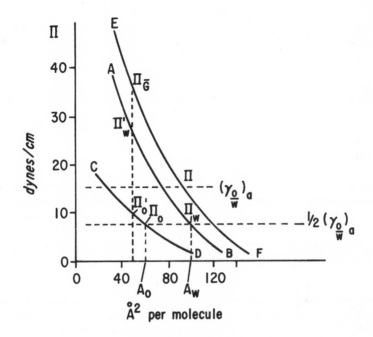

Fig. 2. Curves of Π-A of the mixed film of an o/w
microemulsion. Curve AB represents the water side
and CD the oil side; curve EF is the sum of AB and
CD. Because $\Pi_{\bar{G}} > (\gamma_{o/w})_a$, expansion of the film
occurs spontaneously from the original $\Pi_w^!$ and $\Pi_o^!$ at
50 A to the final Π_w and Π_o at A_w and A_o. Curvature
is effected as the ratio of the area/molecule at the
two sides of the film changes from 1/1 to A_w/A_o.
Redrawn from Ref. (32), courtesy Academic Press, Inc.

difference $\Pi_{\bar{G}} - \Pi$ or $\Pi_{\bar{G}} - (\gamma_{o/w})_a = -\gamma_i^!$. In this case, $-\gamma_i^!$
is the negative interfacial tension before curvature.

A similar hypothetical curve for a w/o emulsion was also
made (32). The higher values of A_o at low pressure at the oil
side of the interphase and the low value of A_w at both $\Pi_w^!$ and
Π_w produces a much larger area, $\Pi_w^!, \Pi_o^!, \Pi_w, \Pi_o$, in these systems
than in the o/w ones. Along with the greater difficulty of
providing molecules to fill the water side of an o/w microemulsion than to fill the oil side of a w/o emulsion with oil
molecules, this explains the empirical observation that w/o
microemulsions are much easier to find than o/w ones.

Such considerations led to a generalized representation of
how microemulsions form. This is illustrated in Fig. 3 (33)

without showing how the water or oil sides of the interphase are filled with the molecules that effect its curvature.

This filling of the vacuum (which Nature abhors) has presented a continuing problem. There is good reason to assume that oil molecules of the proper configuration can readily penetrate and remain in place among the tails of the curved interphase irrespective of its ionic or nonionic nature. The problems at the water side are, however, more complex. These will be treated in a preliminary fashion below.

To conclude the present train of thought, when the initial interfacial tensions at either side of the interphase are equal, or nearly so, the system will consist (29) of layer structures (lamellar micelles) in which the oil and water bulk phases are alternatingly dispersed or the system will consist of hexagonal arrays of cylinders of water (or oil). These conditions may occur as water is added to a w/o microemulsion but will, after inversion, disappear, as an o/w microemulsion forms. For the sake of accuracy, in such systems only the isotropic, fluid systems are microemulsions; the anisotropic and usually viscous dispersions are liquid crystalline phases.

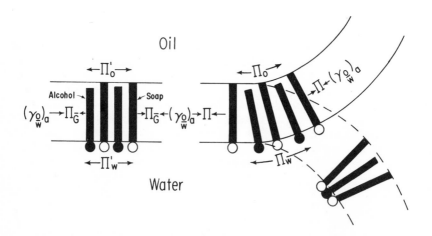

Fig. 3. Diagram illustrating the mechanism of curvature of a microemulsion film. The sum of the pressures at the sides of the flat film is $\Pi_{\overline{G}}$, and the sum of the pressures at the sides of the curved film is Π. The stress of the pressure gradient due to Π_O' and Π_W' is relieved by bending until $\Pi_O = \Pi_W$ or $\Pi = (\gamma_{o/w})_a$. The degree of curvature is dependent on $\Pi_{\overline{G}} - (\gamma_{o/w})_a$. Redrawn from Ref. (33), Fig.4., by courtesy of the Journal of the Society of Cosmetic Chemists.

To return to the water side of the interphase, the molecular interactions responsible for the development of the initial, thermodynamic, two dimensional pressure there are, as at the oil side, not solely determined by the interactions between the surfactant and cosurfactant. Molecules of the bulk water phase, cations (or anions in the case of cationic emulsifiers), polyethylene oxide chains (or other moieties) in the case of nonionics as well as coupling agents (like glycols) are all involved. In some way they combine to fill the voids not filled by the heads of the surfactant and cosurfactant species at the outer side of a sharply curved interphase. Moreover, in the case of elongated species like soap, alkyl sulfonates, long chain alcohols, etc., the leverage available at the polar side of a 25 A thick interphase is usually only 1/5th that at the oil side. The reason is that the water soluble molecular species may only penetrate the interphase to a maximum depth of about 5 A whereas oil molecules may penetrate to a depth of about 20 A. In other words, the mechanical fulcrum is a circle about 4/5th of the way through the interphase. Thus, to effect the same degree of curvature, water moieties are asked to exert five times as much lateral force per unit of length in the interphase to form an o/w microemulsion as are the oil soluble species to form a w/o emulsion. This requires much more specific interactions at the water side than the oil side and may be another explanation for the dearth of good o/w microemulsions.

In spite of these complexities, some progress has been made in how the water side of the interphase is filled with molecules. Although this is only qualitative, it seems to be in the right direction and could form the basis for further discriminating experimentation. This progress is now discussed.

In the area of ionic systems, Sears and Schulman (34) made a relevant finding when they observed that electrostatic repulsion among adjacent soap molecules was not an important factor in determining the degree of expansion of a monolayer. Rather, the expansion of the monolayer was sensitive to the size of the hydrated cation--the larger its size, the greater the expansion. Over a wide range of pH values, they found that differences in the degree of ionization of fatty acids plays only a small role in determining the Π-A relationship in comparison to the specific cation effect. Assuming the diameter of a water molecule to be 2.7 A and that a sheath of water one molecule thick surrounded the cation, areas per cation were obtained as follows: Li^+, 35.7 A^2; Na^+, 42.3 A^2; and K^+, 51.5 A^2. This certainly explains why Rodawald found potassium an essential ingredient in his Carnauba wax emulsions. It gave him a little more leverage at the water side of the interphase. This concept also quashes the idea that inversion of w/o microemulsions is dependent on ionization of the surfactant. Obviously, interactions other than ionic ones are involved.

Unfortunately, the idea that the surface pressure of the alkali metal cations increase in the order $Li^+ < Na^+ < K^+$ does not appear to hold for all anionic surfactants. Weil (35) found the order to be reversed for the alkyl sulfates. This indicates that although the size of the cation influences the expansion of the oil/water interphase, this expansion does not depend solely on the size of the cation; the structure of the tail also appears to be involved.

Some cation behavior, however, seems to be consistent if we consider the emulsifying efficiency of cations in terms of chaotropy, after the nomenclature employed in immunology (36). This was suggested to the writer by Alec Bangham, Institute of Animal Physiology, Babraham, Cambridge, England. By writing the Kekule formula of AMP as Schulman and Montagne (24) did, it seems obvious that the two symmetrical methyl side chains are responsible for a major lowering of the molecule's charge density, thus increasing its chaos-forming or disordering ability. This can disrupt the short range forces in the water phase. This combined with the 2:1 stoichiometric ratio of cation to anion would seem to be an important combination in expanding the water side of the interphase.

This brings us to the bulk water phase itself. There is reason to believe that its molecules behave differently adjacent to a highly oriented vapor phase monolayer than they do in the bulk phase and that such behavior has an effect on the formation of microemulsions. Some attention will be given to this matter without offering any firm hypothesis.

Bulk liquids, particularly water or hydrogen bonding liquids like it, display a tendency towards an orderly rather than a random arrangement of its molecules. In recent years this image of the structure of water has been embodied in a number of models, of which the ice-type lattice is one (37). Much earlier, Hardy (38) suggested that this tendency to organize might be reinforced in depth in proximity to a highly oriented monolayer. McBain (39) imaginatively visualized that the relayed action of intrinsically short range van der Waals-London forces operate in the same way that a magnet can pick up a long chain of iron filings far beyond that at which the attraction becomes negligible. Finally, Davies and Rideal (40) presented evidence that several layers of water adjacent to a mixed film may be oriented to form a rather rigid layer of "soft ice" of the consistency of butter or toffee. Such a state of affairs does not seem to favor microemulsion formation. Something would seem to be needed to break down this structure in order for water molecules to readily flow into the water side of the highly curved interphase.

To break down this structure, Prince and Sears (41) proposed that the heads of the interphase tenants, and possibly their

associated species, array themselves so as to replace oxygen
and hydrogen atoms in the lattice structure of the adjacent
bulk water phase. Under these conditions, the water phase
"sees" the water side of the interphase as possessing an array
of molecules similar to its own lattice structure so that the
interfacial tension would be zero. Such behavior could explain the "flow" of water molecules among the polar heads of
the tenants. This performance does not necessarily determine
the extent nor direction of curvature of the interphase by itself; this would seem to be also controlled by the interactions among the species at the tail side of the interphase.

Some support for this general idea was offered by Fowkes
(42) when he suggested that film pressure was a function of
the chemical potential of water molecules, that water molecules must be area determining and be present in films rich in
alcohol, i.e., when the ratio of alcohol:detergent is greater
than 1:1. He made this proposal on the basis of the fact that
at the air/water interface of a mixed film of cetyl alcohol
and sodium cetyl sulfate, an abrupt change in the Π-A curve
was observed when the ratio of alcohol:detergent reached 1:1.
Moreover, as the alcohol ratio increased he further proposed
that the water molecules were driven out of the film again until they disappeared completely at 100% alcohol.

This and other data led Prince and Sears to suggest that
alcohol molecules and the associated inclusion of oil molecules into the interphase, not only disorder the interphase
but trigger the collapse of the water structure adjacent to
it. In this context, a macroemulsion could be differentiated
from a microemulsion by the degree of water structure adjacent
to the interphase, the higher degree of structure being associated with the macroemulsion.

The foregoing is consistent with the view that a system of
oil, water, surfactant, and cosurfactant (as defined above)
may consist of two bulk liquid phases, one dispersed in the
other in the form of microdroplets by virtue of the forces
exerted by an interphase or third phase. The direction and
magnitude of the curvature of these droplets is, in turn, imposed by the influx of bulk phase molecules which flow into
the oil and water side of the interphase so as to equalize the
chemical potential among the three phases. Whether or not
microdroplets form depends upon the amount and kind of molecules comprising the components of the system as well as the
amount of water. If no curvature forms, the system is liquid
crystalline. This, basically, was Schulman's thesis.

A certain amount of support has been given to this thesis
by the results of high resolution nuclear magnetic resonance
(NMR) studies. Cooke and Schulman (29) were the first to apply this new technique to the study of microemulsions. Their
results indicated that the surface of the droplet, i.e., the

interphase, in a water-in-benzene microemulsion, was liquid
but that the chains of the soap were somewhat constrained.
Hansen (43), studying the same system, found that the polar
ends of the oleate molecules were relatively immobilized at
the water side of the interphase but that the terminal methyl
groups were free to reorient in the benzene phase. Hansen
also found that the alcohol (hexanol) exhibited no motional
restriction and presumably was free to exchange rapidly between the interphase and benzene phase. The water in this
system exists in two conditions: (1) an approximately 1 A
thick layer of water molecules of low mobility associated with
the surfactant polar heads (presumably hydrogen bonded there-
to) and (2) a core of water similar in mobility to a bulk
water phase. The exchange of water between these sites is
rapid. One is reminded of the statement of Ekwall, Mandell,
and Fontell (44) in which they indicated that when all the
hydrogen bonding sites were occupied (on the polar groups of
the surfactant) then water can be considered as a dispersion
medium in the true sense, with the ordinary properties of
water. Hansen was not unmindful of this. Such a bulk water
phase dispersed in benzene points to the fact that these systems are emulsions, not single phase micellar solutions.

Earlier, Gillberg, Lehtinen, and Friberg (45) had utilized
NMR and IR techniques to elucidate the structure of phase
equilibria diagrams of hydrocarbon, water, surfactant, and
soap. By considering that hydrocarbon molecules penetrated
the tails of the micellar solution (L_2 phase), they deduced
that the resultant lower packing density among the tails enhanced the solubilization of water in the aggregate. On the
other hand, at very high concentration of hydrocarbon, the
transport of alcohol from the micelles (or interphase) to the
intermicellar solution (bulk oil phase) was strongly increased, reducing the ratio of alcohol:soap in the interphase.
The NMR (along with the IR) data indicated that the hexanol
did not aggregate in the intermicellar region (bulk phase).
This led these authors to conclude that it is the distribution
of hexanol between the interphase and bulk oil phase which
determines the water solubilizing power of the solutions.
This was previously expressed in the thermodynamic term
$(\gamma_{o/w})_a$ (30). Hansen expressed the overriding thesis rather
circumspectly: "Thus, microemulsions cannot be thought of
simply as pure oil and water regions separated by static
surfactant-containing interfacial films. The existence of a
surfactant at the oil/water interface and a cosurfactant which
distributes itself between the interfacial region and the oil
phase may be a general requirement for microemulsion stability
and it explains why two surfactant species are necessary for
microemulsion formation."

Finally, Zlokower and Schulman (46) confirmed that the

immobilization of water at the water side of the interphase is
the consequence of its participation in a hydrogen bonded network. However, as opposed to the positioning of this network
by Schulman and Montagne (24) at the <u>outer</u> side of the AMP
species, many an Angstrom unit away from the actual interface,
it was now proposed that the hydrogen bonded network lies in
the same plane as the polar heads of the film tenants. Although this supports some of the foregoing thinking in this
area, it also points up some of the uncertainties still associated with the interaction between water and the water side
of the interphase regardless of whether the system is w/o or
o/w.

Before closing the case for the Schulman school of thought,
it is illuminating to review how Schulman over the years faced
the issue of microemulsion versus swollen micelle. In Bowcott
and Schulman (5), p. 285, he noted that the assumption that
all the potassium oleate was in the interface in every system,
was limited. When there is not enough water present to provide a large enough interfacial area for all the soap to reach
the interface, the clear system foamed readily owing to the
soap <u>not</u> in the interphase. On the basis of previous work,
Schulman estimated that 100 A was about the smallest size of
water droplet that can be formed in these systems. In the
electron microscope work (3) in Fig. 13, oil droplets of 75 A
and 150 A in diameter were shown at 280,000x magnification.
In the ultracentrifuge, these two sizes appeared as sedimentation zones. Schulman felt that the 75 A droplet approxi-
mated the dimensions of a swollen micelle, agreeing with the
low angle X-ray pictures taken in 1948 (14). They were also
of the same order of magnitude as the myelinic figures illustrated by Stoeckenius in the first part of the paper. A summation came in Cooke and Schulman (29), Schulman's last paper
on microemulsions, when it was stated that the molecular
weight of the dispersed phase of a microemulsion was greater
than that of most micelles containing solubilized materials.
Thus, high ratios of solubilized to solubilizing materials can
be obtained. These systems were stable. The volume of oil to
water could vary over wide ranges in both the o/w and w/o systems if proper amphipathic agents were employed.

This theme of differentiating microemulsions from micelles
on the basis of aggregate size was recently revived by Prince
(10). When the interphase is 25 A thick, there is an abrupt
change in the ratio of the volume of the interphase to the
volume of the core in the neighborhood of a droplet diameter
of 100 A. Below this diameter the degree of aggregation appears to be determined by the rules governing the formation
of micelles. Above this diameter, curvature is determined by
specific interactions among oil molecules and surfactant tails
as well as by the interactions between surfactant heads and

water in a well-defined interphase. It is the possibility of
such an initial, duplex film, capable of possessing different
tensions at each of its sides that distinguishes the microemulsion from the micellar solution. A tension gradient across
across a freshly formed interphase is able to control its curvature, and thus droplet size, over a much wider range than
can the interaction between surfactant heads and water as in
the case of micelles.

Prince went one step further. Based on the data furnished
by Shah and Hamlin (47), by Bowcott and Schulman (5), and by
Cooke and Schulman (29), a hypothetical phase map was drawn,
Fig. 4, qualitatively defining the regions of microemulsions,
micellar solutions, liquid crystalline phase (mesophases), and
macroemulsions. In this diagram, the criterion used to distinguish microemulsions from micellar solutions was a diameter
of 100 A. The diagram differed from most "phase equilibria
diagrams" of micellar solutions in that the regions of microemulsions and liquid crystalline phases bore a much more rigid
relationship to one another.

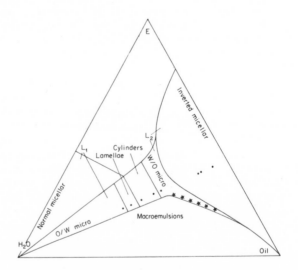

*Fig. 4. Hypothetical phase map showing regions of
microemulsions and micellar solutions. o are points
of Ref. (47); * are points of Ref. (5). Reproduced
from Ref. (10), courtesy Academic Press, Inc.*

V.   COMPLEMENTARY STUDIES

Schulman's investigation of microemulsions, although in
depth, were basically of a preliminary nature. It is therefore not unexpected that several spin-offs from his original
work have made their appearance. These have extended and expanded the scope of Schulman's hypotheses, furnishing a

broader insight into microemulsion systems and, in some cases, confirming his inferences. The common denominator of this work was that the dispersions were considered to be formed by means of the molecular forces originating in a mixed film.

Most of these investigations were carried out by Schulman's students or colleagues or by those directly associated with them. Shah was one of Schulman's last Ph.D. students; Dreher, Rosano, Rosoff, and Sears were post doctorals for a number of years; Robbins was associated with him via industrial consultation; and Giniger and Kai Li Ko(Yuan) were Ph.D. students of Rosoff and Sears, respectively. Dr. A. G. Parts was a colleague in Australia of the late A. E. Alexander, who had begun his emulsion work at Cambridge during the period of the Schulman and Cockbain papers, being, in fact, the author with Schulman of Part 111 of that famous series.

A. Shah's Work

Shah and Hamlin (47) made a series of dispersions of hexadecane, hexanol, and potassium oleate with increasing amounts of water. At low ratios of water to hexadecane, the system was fluid, clear, and transparent, cf. Fig. 1 of Chapter 1. At a ratio of water to oil below 0.1, molecular solubilization of water in the system was presumed to take place. Between a ratio of water to oil of 0.1 to 0.6, the system was considered to be a w/o microemulsion. The existence of these two types of aggregates was supported by the resistance measurements as well as by an upfield shift in NMR spectra. Further addition of water caused the system to become turbid and birefringent, corresponding to cylinders of water and then lamellae. At a ratio of water to oil of about 1.3, the system again became clear and fluid, indicating that it had completely inverted to an o/w microemulsion. The transitions from clear to turbid to clear were again supported by NMR data. This was the first time that this sequence had been demonstrated experimentally, although Schulman, Matalon, and Cohen (18) had demonstrated the existence of cylinders and lamellae in translucent non-ionic dispersions. It is noteworthy that the lamellae converted into o/w microemulsions without passing through a stage consisting of cylinders of oil--a situation which would have made the phase inversion symmetrical and would have been more in line with Winsor's proposals (48).

Subsequently, Shah *et al.* (49) suggested that the formation of microemulsions resulted from interfacial stability which, in turn, was caused by spontaneous, transient, negative interfacial tension resulting from interactions between surfactant and cosurfactant. It was assumed that the original interfacial tension was reduced by the soap and not the alcohol as suggested by Prince (30). On the basis of the electrical measurements, it was also proposed that these microemulsions were true dispersions of one liquid in another and were not cosolubilized systems.

Falco, Walker, and Shah (50) extended the study of the hexadecane, hexanol, potassium oleate, and water systems by measuring their rheological properties. Striking increases in viscosity were found to correspond to the viscoelastic stages in which cylinders and lamellae existed. It was also found that the viscosity in the lamellar, liquid crystalline region increased initially and then leveled off with an increase in shearing time. This was presumed to result from disordering and entanglement of the lamellar aggregates. Further work in this area utilizing X-ray scattering and freeze-etching electronmicroscopy (51) indicated that before agitation the lamellae were oriented parallel to one another. On shaking, the arrangement of the lamellae became disordered and a significant breakdown of the lamellae took place.

Shah (52) and the authors of ref. (51) also proposed that isotropic, transparent, and stable systems may be microemulsions or cosolubilized systems (in which the surfactant and cosurfactant form a mixture which can molecularly disperse (dissolve) both oil and water). By means of electrical conductivity, NMR, spin relaxation times, and rheological measurements, they differentiated between the microstructure of systems which differed only in the chain length of the cosurfactant, i.e., between hexanol and pentanol. Upon dilution with water, the oil, surfactant, and cosurfactant systems both became birefringent and translucent at an intermediate range of water content. The electrical conductivity of the pentanol system, however, underwent a continuous decrease instead of fluctuating. Spherical droplets of water were conspicuous by their absence. Also the liquid crystalline gel stage occurred at a lower water ratio in the pentanol system. The hexanol system, on the other hand, contained spheres of water and of oil on either side of the gel stage.

Earlier Shah had made a very perspicacious observation (53). Careful review of the work on a variety of mixed surfactant systems led him to conclude that the striking changes in the properties of these systems at the 1:3 ratio of surfactant to cosurfactant were due to a common denominator. This, he proposed, was the two-dimensional hexagonal packing of these species resulting in closer molecular packing and a higher degree of stability of the mixed films at interfaces. In this hexagonal packing arrangement, molecules of one type occupy the corners and those of the other type occupy the centers of the hexagons in the adjacent plane.

B. Rosano's Work

Rosano, Peiser, and Eydt (54) extended the work of Bowcott

and Schulman (5) and Cooke and Schulman (29), employing both soap and lauryl sulfates as the surfactants. Microemulsion formation was insensitive to the nature of the cation with the sulfate but quite cation dependent in the case of the soaps. These authors also observed that the addition of water does not increase the total interfacial area of w/o microemulsions. Utilizing Adamson's proposal (55), they suggested that the counterbalancing of the Laplace and osmotic pressures in these systems accounted for the experimental observation that w/o microemulsions are easier to form than o/w types.

Gerbacia and Rosano (56) studied the influence on w/o microemulsions of varying the chain length and type of cation of the surfactant as well as the nature of the surfactant (soap vs. alkyl sulfate). Hexadecane and benzene were used as oil phases. The cosurfactant was pentanol. Based upon the distribution of pentanol between the dispersed phase and the interphase, calculations of ΔG were made. These are discussed later in the section on "Thermodynamics," VI,B. The effect of changing the chain length of the surfactant and the nature of the solvent on the amount of pentanol in the microemulsion was noted. The effect of changing the size of the cation was not as clear cut.

These investigators also conducted experiments which seemed to indicate that it is possible for the interfacial tension of a microemulsion to fall to zero for a short time owing to the redistribution of the pentanol while the equilibrium interfacial tension remains positive. The method or order of mixing of the components of the system seemed to be the key to this observation. The diffusion process seemed to be a necessary condition for spontaneous emulsification but was not an essential condition for microemulsification.

This work did not appear to take into consideration the effect of cosurfactant chain length upon the structure of the solvent nor of the fact that pentanol systems are suspect as cosurfactants for microemulsification (5, 30). A longer chain alcohol might have given much more meaningful results.

In this connection, Shah et al. (51) suggested that some of the microemulsion systems reported in the literature relating to tertiary oil recovery and which exhibited low electrical resistance at low water:oil ratios, may be cosolubilized systems rather than microemulsions because of the solubility of the alcohol in water. The data of Gerbacia and Rosano are subject to the same criticism in connection with their hexadecane systems. Their results with benzene had been interpreted by Bowcott and Schulman (5) along similar lines. Using pentanol and butanol, Bowcott and Schulman (p. 288) found that although a direct proportionality existed between the total alcohol and benzene, indicating that they were w/o systems, both systems had high conductivities until well diluted with benzene. This suggested to them that the continuous phase, although predominantly benzene, might have been a ternary system containing water.

It is of interest that in a subsequent review paper Rosano (57) plotted a ternary phase equilibria diagram of 0.3M aqueous potassium oleate, n-hexadecane, and 1-hexanol in which he found two microemulsion systems, o/w and w/o, separated by a clear birefringent gel system. This is in qualitative accord with the phase map offered by Prince (10).

In this same paper, ignoring not only the possibility of interactions between the tails of the surfactant and cosurfactant species and the oil phase but also the concept of negative interfacial tensions which such interaction engendered, Rosano presented a theory for the development of temporary zero interfacial tension based on the interfacial diffusion of the surfactant and cosurfactant, in accordance with his paper with Gerbacia. However, he did not exclude the factors of film penetration, "complexing," and zero interfacial tension which Schulman had proposed.

Rosano and Weiss (58) titrated mixtures of oil, water, and cosurfactant to clarity with concentrated solutions of surfactant. In this way they hoped to find the optimum film combination which produced a transitory condition of zero interfacial tension while o/w microemulsions were being formed. It was demonstrated that a redistribution of the surfactants among the several phases was necessary for microemulsification. They concluded that the systems were kinetically rather than thermodynamically stable.

C. Robbins's Work

A theory for the phase behavior of microemulsions has been developed by Robbins (59) which is consistent with the concept that interactions in a mixed film are responsible for the direction and extent of curvature and, thus, of the type and size of the droplets of microemulsions. In his most recent model for microemulsions stabilized with nonionics, the heads and tails of the interfacial species are seen as acting as separate uniform liquid phases, with water dissolved in the heads and oil dissolved in the tails. The kind and degree of curvature is imposed by the differential tendency of water to swell the heads and oil to swell the tails.

It is refreshing in this day of a disposition to neglect previous investigations in the field as irrelevant, to find such a thorough and accurate review of previous work. Robbins's references are a paragon of excellence and serve as quite a complete source of background material for microemulsions or micellar solutions, particularly that based on the work of Winsor.

Robbins's model, Fig. 5, quantitatively predicted phase behavior in oil, water, and surfactant systems. He proposed a lateral stress gradient resulting from differences in the swelling of the heads and tails across the interface. This stress gradient was expressed in terms of physically measurable quantities: surfactant molecular volume, interfacial tension, and interfacial compressibility (defined as the fractional change in molecular area with interfacial pressure).

In his earlier work, the stress gradient was viewed as being responsible for the direction of film curvature and the "Laplace" pressure difference across the curved interface. Relating the pressure difference to the fugacity of water in a w/o microemulsion, Robbins established thermodynamic criteria for spontaneous water uptake without postulating a negative interfacial tension.

As a basis for his theory, he has utilized equations generated by the geometry of two concentric spherical shells surrounding the droplet and by the force balance across these shells. Equations are developed for both o/w and w/o droplets which relate phase behavior to interfacial tension, the volume of the surfactant heads and tails, and interfacial compressibility.

Robbins's theory correlates water and oil uptake with the idealized ternary diagram and Winsor's transitions in micellar types. Where water and oil uptake are known, interfacial tension can be predicted and vice versa. The theory also predicts droplet size and interfacial concentration of adsorbed surfactants in terms of the number of molecules per droplet.

D. Other Work

Elbing and Parts (60) found that when boiling water is added to a solution of vinyl stearate in a nonionic surfactant (Renex 690), an opalescent o/w emulsion formed on cooling. Light scattering measurements confirmed that the droplets were in the microemulsion range. It was proposed that the formation of these emulsions involved solution of the oil in the soap-rich phase above the cloud point, followed by dispersion in the aqueous phase on cooling. The solubility of the oil in the nonionic seemed to be a prerequisite for microemulsion formation.

These systems differed from Schulman's by their manner of formation (no viscoelastic gel stage) and by their diminished stability with time. It was the object of this work to provide emulsions of sufficiently small droplet size to permit light scattering measurements to be made during the course of emulsion polymerization, and, in this way, to study the kinetics of the reaction.

Dreher and Syndansk (61) noted that more than one analytical technique was required to distinguish between the disperse and continuous phase in micellar solutions (microemulsions) used in oil recovery and well stimulation. They recommended conductivity and miscibility measurements, in combination, to obtain a clear indication as to whether oil or water is the external phase even at extreme concentrations of the internal phase. Their studies also showed that as much as 70 weight percent of the total system can exist as the internal or disperse phase in an optically clear system.

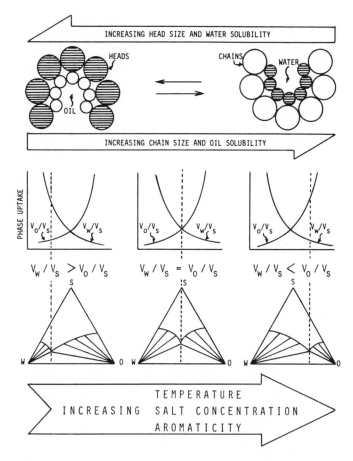

Fig. 5. Microemulsion theory. Reproduced from Ref. (59), courtesy of the Society of Petroleum Engineers of A.I.M.E.

Aiachini, Bonadeo and Lavazza (62) addressed themselves to the technology of microemulsions and solubilized systems as applied to the cosmetic and pharmaceutical industries. They presented an interesting theory based on a quantitative relationship between dispersive and dispersed phases well illustrated by means of simple diagrams. Cosolubilizers were discussed in practical terms. The emphasis was on "hydro-dispersible perfume oils" and the balancing of the formulae not only to obtain optimum perfumery effect but to minimize the risk of chemical and structural perturbations. Disper-sions of vitamins, hormones, and other fat-soluble active sub-stances were included within their purview. However, they erroneously put ringing gels in the category of microemulsions instead of liquid crystalline phases.

Kai Li Ko (63), under the direction of Professor D. F. Sears, re-examined the work of Bowcott and Schulman (5) and Cooke and Schulman (29) in the context of light scattering measurements, using lithium, sodium, and potassium oleates with varying amounts of water as well as different alcohols and hydrocarbons. Unlike previous work by Schulman and co-workers (16,29), who based their calculations on single tur-bidity measurements, this investigator calculated her results from H_c/τ plots of varying concentrations of the microemulsion systems.

Both w/o and o/w systems were studied. In the w/o disper-sions a lower range of water content was employed than in the Bowcott and Schulman work. Although the direct determination of the ratio of alcohol to soap in the interphase of o/w sys-tems is not obtainable as in the w/o systems, these ratios can be obtained indirectly by the results of light scattering. This was done, as was the calculation of the diameter, 172 A, for o/w microemulsions stabilized with potassium oleate and 200 A, for systems stabilized with sodium oleate. As in the w/o systems, the molecular weight and radius of the droplets are larger for the sodium system but the area per soap mole-cule and total interfacial area are larger for the potassium system.

One of the most interesting outgrowths of this light scat-tering study was an attempt to quantitatively calculate the size of the w/o droplets or micelles from the composition of the interphase. It was assumed that the light is scattered from the outer edge of the hydrated cations and that free water molecules exist inside the core. The volume of the wa-ter molecules was taken at 29.89 A^3 and the minimum areas of the lithium, sodium, and potassium cations at 44.4, 35.02, and 31.17 square Angstrom units, respectively. Kai Li Ko inferred that when no water molecules were inside the droplets, the system was a solubilized one. She compared this to Shah's conclusions from conductivity measurements (47). Also, in the o/w systems she proposed that the total interfacial area was determined by the number of closely packed cations at the water side of the interphase.

Molecular interactions in the interphase were based on the assumption that a hexagonal array of alcohol and soap mole-cules produced the closest possible packing in two dimensions.

Planar configuration diagrams were drawn for alcohol:soap
ratios of 3:1, 2:1, and 1:1. When more than three alcohol
molecules per soap molecule were present in the interphase,
the optimum packing was destroyed, indicating that this was
the limiting structure for maximum solubilization of water.
This 3:1 ratio coincided with Shah's proposal (53) for maximum
stabilization. Kai Li Ko also observed that the 1:1 ratio
produced a rectangular instead of an hexagonal array and was
just capable of disordering the interphase enough to yield a
transparent system. This would appear to have some theoretical significance when considered with Fowkes's work (42).

If the cations were surrounded by a layer of water one
molecule thick, they possessed the largest area when the alco-
hol:soap layer in the interphase was 3:1 and the lowest, when
the ratio was 1:1. This explained why the limiting amount of
water for the micelles stabilized with lithium soap is higher
than that for sodium and potassium. In other words, the effect of decreasing the area of the hydrated cation has the
same effect on micro-droplets as increasing the chain length
of the alcohol. It was also confirmed that w/o systems stabilized with higher alcohols are more restricted as to the
amount of water they can solubilize. Moreover, the length of
the alcohol chain had more influence on water retention than
did the type of cation.

Along with Rosano et al. (54), Kai Li Ko found that the addition of water to the w/o systems increased droplet size at
constant interfacial area. She explained this on the grounds
that increasing the radius of the droplets compressed the film
so that less alcohol was required to clear the system.

Giniger (64), under the direction of Professor M. Rosoff,
attempted to reconcile the properties of microemulsions prepared by Schulman and his co-workers with those shown on the
phase diagrams of Friberg and others. The four-component systems of Schulman were made and equilibrium was approached in
several ways. The effect of temperature, time, and composition was studied using light scattering measurements of un-
polarized and polarized light to detect phase transformations
and the presence of anisotropic elements. Emphasis was
placed upon the micellar properties rather than the aspects of
the systems akin to emulsions.

It was found enlightening to explain the properties of four
component systems by drawing three-dimensional phase diagrams.
Such tetrahedrons were drawn for Schulman systems as well as
for those developed by Friberg and Shinoda.

Giniger pointed out that conventional phase diagrams may
not reveal a complete picture of the system. It was shown,
for example, that some systems which were designated isotropic
by birefringence measurements contained anisotropic components
when examined by light scattering depolarization measurements.

He also observed that an isotropic phase can contain aggregates of different sizes, shapes, and structures. He suggested further that some systems of micellar micelles may possess anisotropy and that micellar clusters may form when a phase boundary is approached.

It was also determined that the finding of Gerbacia and Rosano (56) that microemulsification depended upon the order of mixing was incorrect. By ultrasonication, Giniger made microemulsions of their systems. From this he concluded that whether microemulsions occurred spontaneously or nonspontaneously, their formation depended upon intermediate metastable states.

VI. NEW APPROACHES

In addition to the complementary studies, new techniques as well as studies in related fields have added to the knowledge of microemulsions. These have involved conventional and nonconventional monolayer studies and the rigorous application of thermodynamics to these oil, water, and surfactant systems.

A. Monolayer Studies

Investigation of the components of biological membranes has developed evidence to support the contention that oil molecules and the tails of adsorbed, oriented molecules interact to influence the behavior of an oil and water system. Biologists have known that monolayers of distearoyl lecithin at a hydrocarbon/water interface undergo phase separation which depends on temperature and the length of the fatty acid chains of the lecithin. From studies of these systems, Jackson and Hue (65) concluded that "Any model for the behavior of phospholipid monolayers at hydrocarbon/water interfaces must account not only for the structure of the phospholipid but for the influence of the medium in which the phospholipid hydrocarbon chains are immersed." Apparently, the work of compression required to bring the monomolecular film to the state at which phase separation begins, depends markedly on the nature of the hydrocarbon solvent. Applied to microemulsions, in which the state of the monolayer is fixed at vapor condensed (2), whether or not a hydrocarbon remains in a given interface will depend upon both its structure and that of the tails of the surfactant and cosurfactant species.

In what may turn out to be a very significant approach, the simultaneous use of a film balance and electron spin resonance (ESR) techniques has yielded some novel ideas relating to molecular motion in monolayers. On the basis of their experimental results, McGregor et al. (66) proposed a three dimensional interface. "This concept is associated with the

existence of a transition region between two liquid phases
such that physical properties vary continuously through the
interface from one bulk phase to another. This implies the
existence of at least two environments in which a molecule may
find itself." If one accepts this concept, it follows that
the "surface" concentration function depends upon thickness as
well as upon area. This leads to disturbing conclusions concerning film balance data. In this light, the film balance
measurement is a macroscopic one and is much too gross to dif-
ferentiate between thick and thin interfaces. ESR may thus
prove to be an important tool in probing the constitution of
the microemulsion interphase having different tensions or
pressures at its sides.

B.   Thermodynamics

In general, the treatment of surface chemical behavior by
means of thermodynamic equations suffers from a lack of rigorous relationships between such equations and the molecular
interactions taking place at an interface. Nevertheless in
the past ten years thermodynamic principles have been introduced into the study of microemulsion systems and have yielded
some insight into their behavior. In spite of this, the
phrase "thermodynamic stability" used in connection with microemulsions and equated to zero interfacial tension has not
been fully rationalized and remains to plague the field.
The statement by Schulman and the author that the reduction
in interfacial tensions in these systems was solely the result
of the depression of the original oil/water by the two dimensional surface pressure among interfacial species was naive.
The relationship between zero interfacial tension and thermodynamic stability is much more complex. Among the thermodynamic factors that have been considered as of this writing
are: stress gradients, solubility parameters, interfacial
compressibility, the chemical potentials and concentrations of
all species present in the bulk phases as well as the interphase, enthalpy, entropy, the bending and tensional components
of interfacial free energy, osmotic pressures, and the Laplace
pressure difference across the curved interface. These factors have one feature in common--they treat the interphase as
an oriented monolayer.
The concept of thermodynamically stable microemulsions
evolved via the perceived need for a transient negative interfacial tension to spontaneously form these systems. Based on
Lord Rayleigh's suggestion that lateral pressures in two dimensional films do not develop to any considerable extent
until the adsorbed and oriented molecules almost touch one
another and applying the simple thermodynamic equation used to
explain the reduction in the surface tension of water on a

Langmuir trough, the schematic diagram, Fig. 3, explained in
an unsophisticated way the mechanism of film curvature at a
liquid/liquid interface. In it, molecular interactions among
the film tenants and the film balance equation were combined
to produce the proposition that equilibrium was achieved in
the system when the interfacial tension rose to zero. Implicit
in this thesis was the concept that any displacement of the
equilibrium would again produce negative interfacial tension
which would spontaneously return the system to equilibrium at
zero interfacial tension. Arguments against this model appeared from the beginning.

Adamson (55) presented a model for oil external micellar
emulsions as being equivalent to systems of aqueous micelles
in which Laplace and osmotic pressures are balanced. He observed that a clear phase containing oil, water, surfactant,
and cosurfactant could exist in equilibrium with either a
separate hydrocarbon or separate aqueous layer. Both Winsor
(67) (in anionic systems) and Palit et al. (68) (in cationic
systems) described such clear phases as solubilized systems.
Winsor believed his clear systems to be solutions.

Adamson's point was that a micellar emulsion phase can exist in equilibrium with an essentially noncolloidal aqueous
phase. Winsor (67) and Tosch (69) had found that electrolyte
favored such a state of affairs. Adamson further felt that
the existence of two phase equilibria not only provided "an
unambiguous demonstration" of thermodynamic stability but also
permitted thermodynamic parameters such as activities to be
measured. Tosch et al. (70) undertook such a preliminary
investigation.

In Adamson's model of an ionic w/o micelle, part of the
interfacial free energy derived from the electrical double
layer system of the aqueous interior of the micelle. The free
energy contribution associated with the charge sheet on a
monolayer was treated conveniently (although approximately) in
terms of a Donnan equilibrium. Adamson assumed that the interface had a positive and not extremely low free energy but that
at equilibrium the resulting Laplace pressure was balanced by
the osmotic pressure of Donnan origin. Utilizing an interfacial tension of 5 dynes/cm, the model accounted for the general properties of micellar emulsions and specifically permitted of a quantitative treatment of the effect of electrolyte concentration on the distribution of water and electrolyte between the micellar emulsion and a second aqueous phase.

Adamson's position was not inconsistent with the concept of
negative interfacial tension. In multicomponent systems, several definitions of surface free energy are possible depending
on what constraints are imposed. According to Adamson, "The
ordinary surface tension is the work per square centimeter of
surface increase at constant component activities, whereas in

micellar emulsions, surface area changes are at constant total composition, and the high surfactant content then imposes a considerable concomitant exchange of material between interfacial and bulk regions and consequent change in component activities."

It is of significance that inversion of Adamson's micellar emulsions is not normally expected.

Recently, Ruckenstein and Chi (71) undertook a rigorous thermodynamic treatment of microemulsions to obtain information on stability. Their work indicated that when the free energy change of mixing ΔG_M is negative, spontaneous formation of microemulsions occurs, whereas when ΔG_M is positive, macroemulsions can be produced, which, although thermodynamically unstable, may be kinetically stable. They found that for a specified composition, it is possible to form thermodynamically stable emulsions of both types (o/w and w/o), of one type only, or none at all, depending on the value of the specific surface free energy. Moreover, they were able to account for the size of droplets in thermodynamically stable systems and to predict the occurrence of phase inversion. This work has been discussed in some detail by Friberg in the next chapter.

It is noteworthy that calculations of ΔG were also made by Bowcott and Schulman (5), by Gerbacia and Rosano (56), and by Kai Li Ko (63). Rosano and his co-workers used their calculations of the free energies of adsorption of the cosurfactant to indicate that there is little association between the cosurfactant and surfactant, in agreement with earlier work (54). Bowcott and Schulman observed that the strength of the complex due to the contribution from the nonpolar bonding decreased with shortening chain length and was reflected in lower ΔG values. The negative values obtained by Kai Li Ko indicated a positive adsorption of alcohol from the continuous phase to the interphase.

There is little question now but that the interaction between surfactant and cosurfactant is small and that complex formation as visualized by Schulman in his earlier papers is not responsible for zero interfacial tension or microemulsion formation. The fact that microemulsification only appears to occur at intermediate concentrations of surfactant and cosurfactant recently led to the proposal (10) that for zero interfacial tension to occur, high values of two dimensional pressures normally effected by surfactants adsorbed at the oil/water interface must be counterbalanced by a sharp depression in the original oil/water tension, which, in turn, is brought about by a favorable partitioning of cosurfactant between the oil phase and interphase. Excess surfactant or cosurfactant results in a macroemulsion or some other type of system.

Not a vestige of Schulman's complex concept appears to

remain viable. Even Fowkes's experiment (42) at the air/water interface which seemed to support the inference that at least one mole of cosurfactant per mole of surfactant was needed for zero tension at the liquid/liquid interface has been undermined. Shinoda and Friberg on page 293 of Ref. (72) stated that "If the amount of oil (or water) is less than a critical value, the addition of cosurfactant is not necessary." Their evidence was a micellar solution of benzene in an aqueous solution of sodium lauryl sulfate. If, however, the volume of benzene was increased, a cosurfactant was required.

In this same vein, Friberg (73) has reported the possibility of preparing a single phase emulsion with high water content which is thermodynamically stable. The single phase is a liquid crystalline one, possessing the general properties of an emulsion. It is remarkable in that it is heat-thaw stable in the range of 200°C to -100°C. This is an example of the complexities of the systems in which we are involved.

In a related area, responding to reports that surfactants had been developed which can reduce interfacial tensions between two mutually insoluble liquids to the range of 10^{-4} to 10^{-5} dynes/cm, Reiss (74) investigated the possibility of entropy-induced dispersion of bulk liquids. The spontaneous dispersion of the two insoluble liquids at such low but positive tensions would have a profound effect on the process of tertiary oil recovery. Reiss discovered that spontaneous dispersion can occur below 2 dynes/cm and that below tensions of 5×10^{-4} dynes/cm, small aggregates (microdroplets) are formed. He distinguished spontaneous dispersion due to entropy effects from the formation of micellar solutions due to energetics.

Conditions for the thermodynamic stability of emulsions were considered by Wagner (75). A number of situations were described in which the interfacial tension was zero. Under these circumstances, the Gibbs energy term is small and may be positive or negative. This was observed earlier by Hartley (76) on the basis of qualitative considerations. In addition, Wagner postulated that a thermodynamically stable emulsion can be obtained only if the concentration of surfactant required for zero interfacial tension is lower than its critical micelle concentration. Finally, he notes that interfacial tensions equal to zero can occur only at an intermediate ratio of surfactant and cosurfactant.

C. Mackay's Work

Mackay has observed that the intermediate size of microdroplets--between classical emulsion droplets and micelles--makes them of potential value for studies at the oil/water interface. Conversely, the results of such studies could throw light on the molecular interactions taking place at the microemulsion interface.

Letts and Mackay (77) had investigated the incorporation of divalent metal ions of Cu, Mg, Mn, Zn, and Co by tetraphenylporphine in benzene-water microemulsions stabilized by cyclohexanol and both cationic and anionic surfactants. Subsequently, they studied the effect of triphenylphosphine on the incorporation of aqueous copper (II) by tetraphenylporphine in a benzene-in-water microemulsion containing cyclohexanol and cetyl sulfate as surfactants. Further work is in progress.

REFERENCES

1. Hoar, T. P., and Schulman, J. H., *Nature (London)* 152, 102 (1943).
2. Schulman, J. H., Stoeckenius, W., and Prince, L. M., *J. Phys. Chem.* 63, 1677 (1959).
3. Stoeckenius, W., Schulman, J. H., and Prince, L. M., *Kolloid Z.* 169, 170 (1960).
4. Prince, L. M., *Soap Chem. Specialties* 36, Sept., Oct. (1960).
5. Bowcott, J. E., and Schulman, J. H., *Z. Elektrochem.* 59, Heft 4, 283 (1955).
6. Schulman, J. H., and Cockbain, E. G., *Trans. Faraday Soc.* 36, 551, 661 (1940).
7. Bancroft, W. D., *J. Phys. Chem.* 17, 501 (1913).
8. Clowes, G. H. A., *J. Phys. Chem.* 20, 407 (1916).
9. Boffey, B. J., Collison, R., and Lawrence, A. S. C., *Trans. Faraday Soc.* 55, 654 (1959).
10. Prince, L. M., *J. Colloid Interface Sci.* 52, 182 (1975).
11. Schulman, J. H., and Friend, J. A., *Soc. Cosmetic Chemists 1*, 381 (1949).
12. Schulman, J. H., and McRoberts, T. S., *Trans. Faraday Soc.* 42B, 165 (1946).
13. Schulman, J. H., McRoberts, T. S., and Riley, D. P., *J. Physiology 107* (1948).
14. Schulman, J. H., and Riley, D. P., *J. Colloid Sci.* 3, 383 (1948).
15. Pink, R. C., *Discuss. Faraday Soc.* 8, 170 (1946).
16. Schulman, J. H., and Friend, J. A., *J. Colloid Sci.* 4, 497 (1949).
17. Winsor, P. A., *Trans. Faraday Soc.* 44, 376 (1948).
18. Schulman, J. H., Matalon, R., and Cohen, M., *Disc. Faraday Soc.* 11, 117 (1951).
19. Goddard, E. G., and Schulman, J. H., *J. Colloid Sci.* 8, 309, 329 (1953).
20. Adamson, A. W., "Physical Chemistry of Surfaces," p. 129, Interscience, New York, 1966.

21. Zisman, W. A., *J. Chem. Phys. 9*, 789 (1941).
22. Prince, L. M., U.S. Patent 2,441,842.
23. Robbins, M. L., and La Mer, V. K., *J. Phys. Chem. 62*, 1291 (1958).
24. Schulman, J. H., and Montagne, J. B., *Ann. N.Y. Acad. Sci. 92*, Art. 2, 366-371 (1961).
25. Prince, L. M., *in* "Emulsions and Emulsion Technology" (K. J. Lissant, Ed.), Part 1, pp. 164-165, Marcel Dekker, New York, 1974.
26. Prince, L. M., *Cosmetics and Perfumery 89*, 47 (1974).
27. Ryer, F. V., *Oil and Soap 23*, 310 (1946).
28. Bowcott, J. E., Ph.D. Dissertation, Cambridge, 1957.
29. Cooke, C. E., Jr., and Schulman, J. H., "Surface Chemistry," pp. 231-251, Munksgaard, Copenhagen, 1965.
30. Prince, L. M., *J. Colloid Interface Sci. 23*, 165 (1967).
31. Prince, L. M., reference (25), p. 144.
32. Prince, L. M., *J. Colloid Interface Sci. 29*, 216 (1969).
33. Prince, L. M., *J. Soc. Cosmetic Chemists 21*, 193 (1970).
34. Sears, D. F., and Schulman, J. H., *J. Phys. Chem. 68*, 3529 (1964).
35. Weil, I., *J. Phys. Chem. 70*, 133 (1966).
36. Dandliker, W. B., and de Saussure, V. A., *in* "The Chemistry of Biosurfaces" (M. L. Hair, Ed.), Vol. 1, Marcel Dekker, New York, 1971.
37. Narten, A. H., Danforth, M. D., and Levy, H. A., *Disc. Faraday Soc. 43*, 97 (1967).
38. Hardy, W. B., *Proc. Roy. Soc. 86A*, 631 (1912).
39. McBain, J. W., "Colloid Science," p. 69, D. C. Heath & Company, Boston, 1950.
40. Davies, J. T., and Rideal, E. K., "Interfacial Phenomena," p. 369, Academic Press, New York, 2nd Ed., 1963.
41. Prince, L. M., and Sears, D. F., "A Theory of Micro-emulsions," paper presented at Microemulsion Symposium of American Chemical Society, Washington, D. C., September, 1971.
42. Fowkes, F. M., *J. Phys. Chem. 67*, 1982 (1963).
43. Hansen, J. R., *J. Phys. Chem. 78*, No. 3, 256 (1974).
44. Ekwall, P., Mandell, L., and Fontell, K., *J. Colloid Interface Sci. 33*, 215 (1970).
45. Gillberg, G., Lehtinen, H., and Friberg, S., *J. Colloid Interface Sci. 33*, 40 (1970).
46. Zlochower, I. A., and Schulman, J. H., *J. Colloid Interface Sci. 24*, 115 (1967).
47. Shah, D. O., and Hamlin, R. M., Jr., *Science 171*, 483 (1971).
48. Winsor, P. A., *Chem. Reviews 68*, 1 (1968).
49. Shah, D. O., Tamjeedi, A., Falco, J. W., and Walker, R. D., Jr., *AIChE J. 18*, 1116 (1972).

50. Falco, J. W., Walker, R. D., Jr., and Shah, D. O. *AIChE J. 20,* 510 (1974).

51. Shah, D. O., Walker, R. D., Jr., Hsieh, W. C., Shah, N. J., Dwivedi, S., Nelander, J., Pepinsky, R., and Deamer, D. W., Paper number SPE 5815, presented at Improved Oil Recovery Symposium of the Society of Petroleum Engineers of AIME, Tulsa, Oklahoma, March 22-24, 1976.

52. Shah, D. O., "On Distinguishing Microemulsions from Co-solubilized Systems," presented at 48th National Colloid Symposium, Austin, Texas, June 24-26, 1971.

53. Shah, D. O., *J. Colloid Interface Sci. 37,* 744 (1971).

54. Rosano, H. L., Peiser, R. C., and Eydt, A., *Revue Francaise des Corps Gras 16,* 249 (1969).

55. Adamson, A. W., *J. Colloid Interface Sci. 29,* 261 (1969).

56. Gerbacia, W., and Rosano, H. L., *J. Colloid Interface Sci. 44,* 242 (1973).

57. Rosano, H. L., *J. Soc. Cosmetic Chem. 25,* 609 (1974).

58. Rosano, H. L., and Weiss, A., Kendall Award Symposium Div. Colloid and Surface Chemistry, Centennial Amer. Chem. Soc. Meeting paper No. 27, April 7, 1976, New York, New York.

59. Robbins, M. L., "Theory for the Phase Behavior of Micro-emulsions," Paper No. 5839, presented at the Improved Oil Recovery Symposium of the Society of Petroleum Engineers of AIME, Tulsa, Oklahoma, March 22-24, 1976.

60. Elbing, E., and Parts, A. G., *J. Colloid Interface Sci. 37,* 635 (1971).

61. Dreher, K. D., and Syndansk, R. D., *J. Petroleum Technology Forum,* p. 1437, Dec., 1971.

62. Aiachini, A., Bonadeo, I., and Lavazza, M., Dragoco Report, p. 53, March, 1975.

63. Kai Li Ko (Yuan), Ph.D. Dissertation, December, 1974, Dept. Physics, Tulane University, New Orleans, Louisiana.

64. Giniger, A. A., Ph.D. Dissertation, January, 1975, College of Pharmaceutical Sciences, Columbia University, New York, New York.

65. Jackson, C. M., and Yue, B. Y. J., *in* "Monolayers," Advances in Chemistry Series 144, p. 202, American Chem. Soc., E. D. Goddard, Ed., Washington, D. C., 1975.

66. McGregor, T. R., Cruz, W., Fernander, C. I., and Mann, J. A., Jr., *in* "Monolayers," Advances in Chemistry Series 144, p. 308, Amer. Chem. Soc., E. D. Goddard, Ed., Washington, D. C., 1975.

67. Winsor, P. A. *Trans. Faraday Soc. 44,* 376 (1948); *ibid. 5,* 762 (1950).

68. Palit, S. R., Moghe, V. A., and Biswas, B., *Trans. Faraday Soc. 55,* 463 (1959).

69. Tosch, W. C., "Technology of Micellar Solutions," Paper

1847B, presented at 42nd Annual Fall Meeting of Petroleum Engineers, Houston, Texas, October, 1947.

70. Tosch, W. C., Jones, S. C., and Adamson, A. W., *J. Colloid Interface Sci. 31,* 297 (1969).

71. Ruckenstein, E., and Chi, J. C., *J. Chem. Soc. Faraday Trans. 11,* 1690 (1975).

72. Shinoda, K., and Friberg, S., *Adv. Colloid Interface Sci. 4,* 281 (1975).

73. Friberg, S., *Amer. Perf. Cosmetics 85,* 12, 27 (1970).

74. Reiss, H., *J. Colloid Interface Sci. 53,* 61 (1975).

75. Wagner, C., *Colloid & Polymer Sci. 254,* 400 (1976).

76. Hartley, G. S., *Trans. Faraday Soc. 37,* 130 (1941).

77. Letts, K., and Mackay, R. A., *Inorganic Chemistry 14,* 2990 (1975); *ibid. 14,* 2993 (1975).

Microemulsions and Micellar
Solutions

STIG FRIBERG

Department of Chemistry
University of Missouri-Rolla, Rolla, MO 65401

and

The Swedish Institute for Surface Chemistry
Stockholm, Sweden

I. STABILITY OF MICROEMULSIONS, BASIC FACTORS

Since their technological (1) and scientific introduction (2), the stability of microemulsions has been intensely
discussed using different approaches. The mixed-film theory
(Chapter 5) limited its approach to the interfacial free
energy; it is obviously an oversimplification (3) because other
factors play an equally important role in the stability of
small-particle systems.

Microemulsions are (4-7) liquid dispersions with diameters
in the colloidal range of 20-80 nm. They might be stabilized
with a combination of an ionized surfactant (a soap or an alkyl
sulphate in the model systems) and a hydrophobic co-surfactant,

commonly an alcohol of medium chain length or with a nonionic surfactant of optimum hydrophilic-lipophilic balance. The latter approach will not be treated; the interested reader may consult a recent review article (8) or original contributions (9-13).

Accepting the microemulsions to be dispersions of colloidal size, the model system used by Ruckenstein (14) is suitable for an estimation of the relative importance of different factors which are instrumental in stabilizing the system. The model consists of monodisperse microdroplets, which are randomly distributed in a continuous liquid phase. The factors of importance for their stability are (14,15) in addition to the surface free energy used in the mixed film theory (Chapter 5), are (a) the van der Waals attraction potential between the dispersed droplets; (b) the repulsive potential from the compression of the diffuse electric double layer; and (c) the entropic contribution to the free energy from the space position combinations of the dispersed droplets.

The van der Waals potential has been calculated (14) using the Ninham-Parsegian approach (16), the energy of the electric double layer from a Debye-Huckel distribution (17) and lowest and highest limit for the entropy estimated from geometrical considerations. The results showed the van der Waals potential to be negligible in comparison to the other contributions to the free energy.

For a Ruckenstein model system at 25°C, for o/w conditions, for a volume fraction dispersed phase $\phi_2 = 0.33$ and for a 1:1 electrolyte of concentration 0.01 M the contribution to the free energy from the compression of the diffuse electric double layer will be (14):

$$\Delta G_{e.d.\ell} = \left[2.2 \cdot 10^{-26}\, R^{-5} + 4.9 \cdot 10^{-40}\, R^{-7} + \right.$$
$$(5.1 \cdot 10^{-33}\, R^{-6})\, e^{-2.4 \cdot 10^6 R}\right] \times \left[1.9 \cdot 10^6 R - 1 + \right.$$
$$\left. (1.9 \cdot 10^6 R + 1)\, e^{-3.8 \cdot 10^6 R}\right] \tag{1}$$

The entropic contribution should be between

$$\Delta G_{Entr.L} = T\Delta S_L =$$
$$(1.54 \cdot 10^{-13} + 9.72 \cdot 10^{-15}\, \ln R)\, R^{-3} \tag{2}$$

and

$$\Delta G_{Entr.H} \simeq T\Delta S_H \simeq$$

$$(6.5 \cdot 10^{-13} + 9.72 \cdot 10^{-15} \ln R)R^{-3} \tag{3}$$

A reasonable approximation is given by

$$\Delta G_{Entr.M} = T\Delta S_M =$$

$$(1.69 \cdot 10^{-13} + 9.72 \cdot 10^{-15} \ln R)R^{-3} \tag{4}$$

Accepting the standard states as zero the corresponding interfacial tension may be calculated

$$\gamma = (\Delta G_{Entr.M} - \Delta G_{e.d.\ell.})R/0.99 \tag{5}$$

As observed in Fig. 1, all values are positive by the most conservative estimate (whole line, Eq. (2)); values may be higher depending on the model. The broken line in Fig. 1 is a more reasonable estimate (Eq. (4)); the dotted curve indicates the highest values from Eq. (3). The model is useful since it provides an estimate of the magnitude of different factors of importance affecting stability. It is not the final solution to the problem, however, since some interactions are not treated.

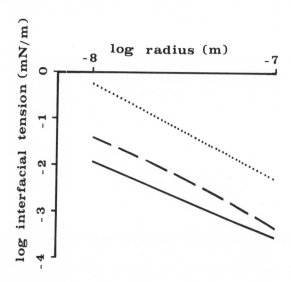

Fig. 1. *Calculations based on Ref. (14) of the interfacial tension for liquid particles in the microemulsion region show positive values:* ——— *Minimum value;* — — — *Reasonable value;* .... *Maximum value.*

The progress of the theoretical treatment of the microemulsion state has recently accelerated, but the statistical-
mechanical treatment necessary to enable *ab initio* calculations still appears to be a thing of the future. Awaiting
this development, empirical relations between microemulsion
compositions and the properties of the components are useful
in order to facilitate the application. A summary of these
relations is given in the next section; a more complete review has recently been published (8).

II.   THREE- AND FOUR-COMPONENT PHASE DIAGRAMS

A.   Three-Component Systems

The three-component phase diagram is a practical tool for
understanding the association phenomena of importance to
microemulsions. The following treatment is concerned with
liquid associations; in addition the liquid crystals are mentioned as related structures. The reader is assumed to be
able to read a three-component phase diagram.*
The associations between the three structure-forming components: water (H_2O), surfactant (S), and cosurfactant (Co-S)
form the basis of microemulsion structure. Their general
behavior is illustrated by Fig. 2, showing four one-phase
areas 1-4. Here 1 and 2 are regions of isotropic solutions;
3 and 4, regions containing liquid crystals. Regions 1-3 are
of interest for microemulsions and will be further discussed.
Region 1 contains molecularly dispersed surfactant at
concentrations below the critical micellization concentration
and normal micelles at higher concentrations, Fig. 2. Hence
the solubility of the cosurfactant is low below the cmc, but
considerably enhanced at higher concentrations owing to solubilization of the cosurfactant in the normal micelles.
Inspection of region 2 shows low solubility of water and
surfactant monomer in the liquid cosurfactant; whilst combinations of the two compounds dissolve to high values. The
presence of water causes an increase of the surfactant solubility from 5 to 30 weight-% and a cosurfactant/surfactant
ratio of 4 increases the water solubility from 5-45 weight-% .

It is essential to observe that the enhanced solubility
of the surfactant is not due to the formation of inverse micelles, which is a common mistake. The area marked C in Fig.
2 contains ion pairs of the surfactant with a few associated
water molecules per ion pair serving to reduce the field

---

*See page 147.

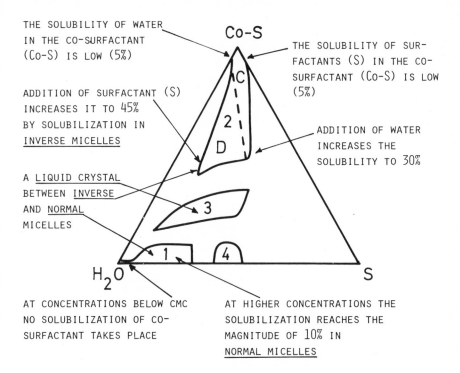

THE SOLUBILITY OF WATER
IN THE CO-SURFACTANT
(Co-S) IS LOW (5%)

Co-S

THE SOLUBILITY OF SUR-
FACTANTS (S) IN THE CO-
SURFACTANT (Co-S) IS LOW
(5%)

ADDITION OF SURFACTANT (S)
INCREASES IT TO 45%
BY SOLUBILIZATION IN
INVERSE MICELLES

ADDITION OF WATER
INCREASES THE
SOLUBILITY TO 30%

A LIQUID CRYSTAL
BETWEEN INVERSE
AND NORMAL
MICELLES

H$_2$O

S

AT CONCENTRATIONS BELOW CMC
NO SOLUBILIZATION OF CO-
SURFACTANT TAKES PLACE

AT HIGHER CONCENTRATIONS THE
SOLUBILIZATION REACHES THE
MAGNITUDE OF 10% IN
NORMAL MICELLES

*Fig. 2.   The main association structures of surfactant
(S) and cosurfactant (Co-S) of importance for microemulsion phenomena are normal micelles (1), ion pairs
(2C), inverse micelles (2D), and liquid crystals (3).*

strength between the two ions. This fact is highly significant for microemulsions; especially in tertiary oil recovery,
where brine is encountered. It is common to call region C a
micellar solution (Chapter 5); this is not correct and from
the application point of view, it is misleading.

Inverse micelles form first where the concentration of
water extends into the area D in Fig. 2. These inverse micelles contain a central core of water surrounded by surfactant and cosurfactant molecules and are dispersed in the liquid cosurfactant. The water solubilizing capacity is strongly
dependent on the surfactant/cosurfactant ratio. Too high a
content of the cosurfactant will cause a separation into two
liquids whilst too much surfactant gives rise to separation of
a different association structure: a liquid crystalline phase.
This liquid crystalline phase has a lamellar structure
as illustrated in Fig. 3. Its properties are completely different from those of the isotropic solutions; its structure is

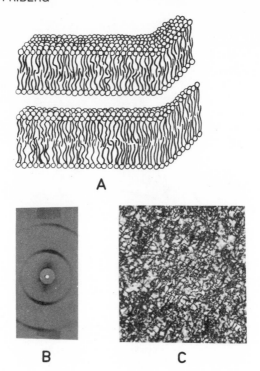

Fig. 3. *The lamellar liquid crystal has a regular structure (A); it gives a characteristic X-ray reflection pattern in the low angle region (B) and is optically anisotropic (C).*

characterized by a long range order organization reflected in its X-ray diffraction pattern in the low-angle area and its optical anisotropy. It is highly viscous and is a natural and expected intermediate in the transition region between normal and inverse micelles. It is essential that it be observed as such and not confused with the microemulsion state.

B.   Four-Component Systems

In addition to the three basic components: water, surfactant, and cosurfactant, microemulsions also contain hydrocarbon, and it is necessary to extend the three-component diagram to a four-component presentation. This is done simply by adding a fourth corner to the triangle in Fig. 4A, forming the tetrahedron according to Fig. 4B. The presentation and discussion of the microemulsion state is a question of how to present solubility areas in this tetrahedron, and there is no doubt that much of the lengthy discussion on microemulsions could have been avoided with a knowledge of these conditions.

If a microemulsion state is experienced it should be observed as an isotropic solution in the marked region in Fig. 4B. The aim of the following presentation is to relate the microemulsion region to the normal and inverse micellar solutions by using such phase diagrams. In order to have some (for microemulsion literature refreshing) connection with reality a model system will be described with the four components (20): water (H_2O), dodecylsulphate ($C_{12}SO_4$), pentanol (C_5OH), and p-xylene (C_6C_2).

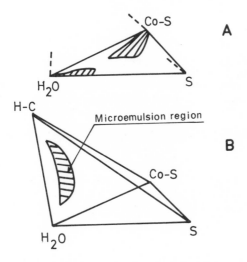

Fig. 4. *In order to obtain microemulsions the two-dimensional phase diagram of the three structure-forming elements (A) has to be extended to a three-dimensional representation according to (B). The microemulsions should contain high amounts both of water and hydrocarbon (the marked region).*

III. W/O MICROEMULSIONS

These are directly related to the inverse micellar solutions, and the rules to obtain them directly follow from the association of the three basic components. Figure 5A shows the basic features of the inverse micellar region of the phase diagram of the structure-forming components (cf. Fig. 2). The isotropic liquid region reaches towards the water corner in a narrow channel with maximum water solubilization for a cosurfactant/surfactant ratio of 3.5. The diagram for 50 weight-% benzene is then introduced as a plane in the tetrahedron according to Fig. 5B. The area for microemulsions is

now observed as a direct continuation of the inverse micellar
area in Fig. 5A. Maximum water solubilization is obtained for
a cosurfactant/surfactant ratio of 3.5 (C)--identical to the
corresponding ratio for the system without hydrocarbon--and
the general features of the solubility area are also similar
to those in Fig. 5A.

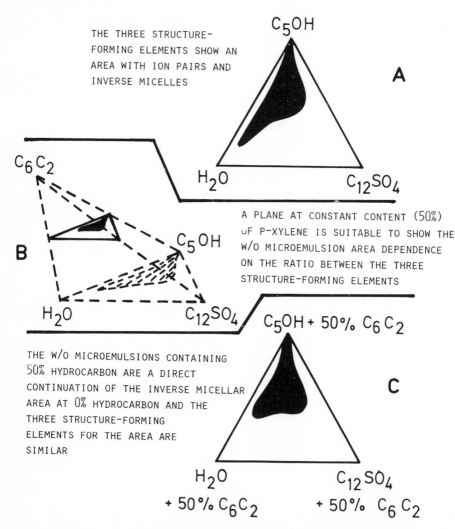

THE THREE STRUCTURE-
FORMING ELEMENTS SHOW AN
AREA WITH ION PAIRS AND
INVERSE MICELLES

C_5OH

A

H_2O $C_{12}SO_4$

C_6C_2

C_5OH

B

H_2O $C_{12}SO_4$

A PLANE AT CONSTANT CONTENT (50%)
OF P-XYLENE IS SUITABLE TO SHOW THE
W/O MICROEMULSION AREA DEPENDENCE
ON THE RATIO BETWEEN THE THREE
STRUCTURE-FORMING ELEMENTS

THE W/O MICROEMULSIONS CONTAINING
50% HYDROCARBON ARE A DIRECT
CONTINUATION OF THE INVERSE MICELLAR
AREA AT 0% HYDROCARBON AND THE
THREE STRUCTURE-FORMING
ELEMENTS FOR THE AREA ARE
SIMILAR

$C_5OH + 50\% \ C_6C_2$

C

H_2O $C_{12}SO_4$
$+ 50\% \ C_6C_2$ $+ 50\% \ C_6C_2$

Fig. 5. The w/o microemulsion region (B) is a direct
continuation of the ion pair and inverse micellar
solution of the three structure-forming elements:
water (H_2O), surfactant ($C_{12}SO_4$), and cosurfactant
(C_5OH). The representation C, the full line part
from B, is convenient for comparing ratios of the
three structure-forming elements (A).

These results show the w/o microemulsions to be inverse
micelles at high water contents and solutions containing
water/ion pair associations at low water contents. It is
essential to observe that this distinction is not a matter of
semantics; the stability tolerance of the ion pair solution
compositions for electrolytes is extremely small.

The corresponding area for higher contents of hydrocarbon
is not included, but at high hydrocarbon content the water
solubilization capacity is extremely small; the solubility
region is critically dependent on the ratio between ionized
surfactant and water (18).

Similar diagrams have been presented for many other systems (18,19); the reader may easily confirm the identity
between the Schulman-Prince w/o microemulsions and these inverse micellar solutions. The term microemulsions serves no
real purpose for these systems; however, since it is well
established and has a certain appeal, it will certainly be
retained. It is, however, important to realize the colloidal
state of these w/o microemulsion systems.

IV. O/W MICROEMULSIONS

The relation between the o/w microemulsions and the micellar solutions is less direct. Figure 6A shows the normal
micellar solubilization of the three basic components water,
surfactant, and cosurfactant. The solubilization of cosurfactant in the aqueous solution is limited. The solubilization of hydrocarbon in the aqueous solution is even more
limited according to Fig. 6B.

When the two are combined the solubilization may be dras-
tically enhanced (20) for aqueous solutions with an optimal
concentration of surfactant, Fig. 6C. The concentration of
surfactant in the aqueous solution is the critical factor;
Fig. 7 reveals the reduction of solubilization when the concentration of surfactant is changed from 15 to 10 or 20%
(W/W). So far very few systematic studies (21,22) comparable
to those of inverse micellar systems have been made on the
o/w microemulsions; the conditions certainly merit scientific
investigation.

V. MIXED FILM THEORY VERSUS MICELLAR ASPECTS

The fruits of the years of investigation using "the mixed
film theory" have been summarized in a "tentative" phase
map, Fig. 4, Chapter 5. It is useful to compare this tentative diagram with one obtained in reality, Fig. 8, in order
to correct defects which may cause problems in the practical
applications of microemulsions.

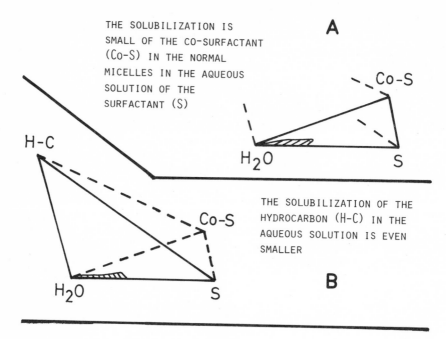

THE SOLUBILIZATION IS
SMALL OF THE CO-SURFACTANT
(Co-S) IN THE NORMAL
MICELLES IN THE AQUEOUS
SOLUTION OF THE
SURFACTANT (S)

A

Co-S

H₂O   S

H-C

THE SOLUBILIZATION OF THE
HYDROCARBON (H-C) IN THE
AQUEOUS SOLUTION IS EVEN
SMALLER

Co-S

H₂O   S

B

FOR AN OPTIMUM CONCENTRA-
TION OF SURFACTANT THE
<u>COMBINED</u> SOLUBILIZATION OF
HYDROCARBON AND THE CO-
SURFACTANT IS LARGE AND A
<u>MICROEMULSION</u> IS FORMED

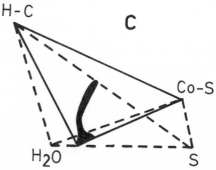

H-C

C

Co-S

H₂O   S

*Fig. 6. The o/w microemulsion region (C), on the
other hand, extends from the aqueous micellar solution for an optimum surfactant concentration in the
aqueous solution.*

According to Fig. 4 , Chapter 5, the combined emulsifiers
are soluble in hydrocarbon. An ionic surfactant emulsifier,
such as potassium oleate or sodium dodecyl sulphate is only
slightly soluble in a hydrocarbon or in combination with a
medium chain length alcohol such as pentanol. This means that
the solubility area along the emulsifier/hydrocarbon axis in
Fig. 4 , Chapter 5, does not exist for soaps such as potassium
oleate. The correct phase diagram, Fig. 8, demonstrates the

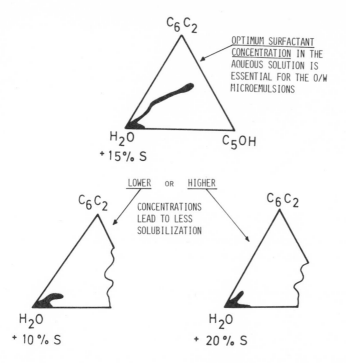

Fig. 7. The importance of optimum concentration of surfactant is proved by the reduction of solubilization at lower or higher concentrations (lower part of figure).

necessity of some water to obtain the solubility of the emulsifier. It should be observed that this is no trifling matter; it is highly significant in the practical application of microemulsions to tertiary oil recovery.

The intermolecular forces between alcohols and soaps are not sufficient to reaggregate the alcohol molecules to provide sufficient shielding of the strongly polar groups of the soap. In order to obtain solubility in hydrocarbons the molecular interaction must be of higher energy provided by the hydrogen bond from the carboxylic group of a carboxylic acid (23) or from a multi-dentate liquid such as glycerol to the carboxylic group of a soap. The carboxylic acid/carboxylic hydrogren bond is obviously stronger than that in a carboxylic acid dimer (∼30kJ/mole) which in turn is stronger than the hydrogen bond in water and in alcohols (∼20 kJ/mole). The soap/carboxylic acid hydrogen bond is sufficiently strong not to be disintegrated at dilution of the compound by a nonpolar solvent (24). The combination soap/carboxylic acid opens up possibilities for preparing microemulsions in which a soap may be completely soluble in the oil phase--an interesting development for tertiary oil recovery.

The second point is in the term "inverse micellar" used to name the part to the right in the diagram, Fig.**4** , Chapter 5. This is not correct, since no micelles are present; the water and surfactant molecules exist as ion pairs in that part of the system. The inverse micelles form first at higher concentrations of water. It is essential to realize the practical implications of this fact; high electrolyte content, such as brine, prevents the formation of microemulsions in the ion-pair area.

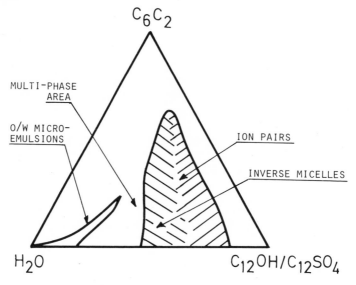

Fig. 8. *Phase regions of microemulsions in the water/oil/microemulsion representation.*

Figure **4** , Chapter 5, presents the o/w microemulsions in a sectorial area emanating from the aqueous corner with a continuous transition into w/o microemulsions. In reality the area for o/w microemulsions is bent and the transition to w/o inverse micellar solutions takes place over multiphase states. These latter frequently involve liquid crystals; unfortunately, that fact is sometimes not observed and the multiphase mixtures are introduced as microemulsions. The mistake is easily made; the systems may appear isotropic and have the appearance of a liquid. The liquid crystal may be dispersed in small spherical aggregates and not observed as such (22). Hence the presence of a liquid crystalline phase is not observed until it starts separating from the liquid after an induction period that may reach several months or years. The distinction in practice between a stable microemulsion and a multiphase system is difficult but essential.

The phase diagrams, such as those in Figs. 5 and 6, are useful for this purpose since they directly display which compositions may be misleading. The multiphase areas including a liquid crystalline phase are found only at the limit of high surfactant/cosurfactant ratios, Figs. 6C and 7A. The instability at high cosurfactant/surfactant ratios is directly observed since the microemulsions will split into two solutions, a phenomenon of immediate reversibility.

VI. PERSPECTIVES

After the recognition of microemulsions as part of micellar solutions and their relations to the concentrations of different compounds, such as in Figs. 5 and 6, a systematic theoretical approach may be rapid and the phase equilibria involving only liquid phases may be clarified within a few years when necessary additional factors to those summarized in the first section of this chapter have been included. The phase equilibria including liquid crystals may be more difficult, however, since their nature of association aggregates of almost infinite extensions causes problems in the form of extremely small differences in free energy between high molecular weight micellar and liquid crystalline associations. In addition low rate constants may be expected, supplementing the theoretical problems with practical ones.

ACKNOWLEDGMENTS

This chapter was prepared during a visit to the Science Centre for Advancement of Postgraduate Studies, Alexandria, Egypt. The author is indebted to the Director of the Centre, Professor El Sadr, and to the UNESCO Chief Technical Adviser, Professor S. N. Srivastava.

REFERENCES

1. Kokatnur, V. R., U.S. Patent 2.111.100 (1935).
2. Hoar, T. P., and Schulman, J. H., *Nature (London) 152*, 102 (1943).
3. Miller, C. A., and Scriven, L. E., *J. Colloid Interface Sci. 33*, 360 (1970).
4. Schulman, J. H., Stockenius, W., and Prince, L. M., *J. Phys. Chem. 63*, 1677 (1959).

5. Shah, D. O., and Hamlin, R. M., *Science 171*, 483 (1971).

6. Prince, L. M., *J. Colloid Interface Sci. 29*, 216 (1969).

7. Cooke, C. E., and Schulman, J. H., *in* "Surface Chemistry," p. 231, Munksgaard, Copenhagen, 1965.

8. Shinoda, K., and Friberg, S., *Adv. Colloid Interface Sci. 4*, 281 (1975).

9. Saito, H., and Shinoda, K., *J. Colloid Interface Sci. 32*, 649 (1970).

10. Shinoda, K., and Ogawa, T., *J. Colloid Interface Sci. 24*, 56 (1967).

11. Becker, P., and Arai, H., *J. Colloid Interface Sci. 27*, 634 (1968).

12. Kitahara, A., Ishikawa, T., and Taniusori, S., *J. Colloid Interface Sci. 23*, 243 (1967).

13. Friberg, S., and Lapczynska, I., *Progr. Colloid & Polymer Sci. 56*, 16 (1975)

14. Ruckenstein, E., and Chi, J. H., *J. Chem. Soc. Faraday Trans. 71*, 1690 (1975).

15. Reiss, H., *J. Colloid Interface Sci. 53*, 61 (1975).

16. Ninham, B. W., and Parsegian, V. A., *Biophys. J. 10*, 646 (1970).

17. Prince, L. M., *J. Colloid Interface Sci. 52*, 182 (1975).

18. Gillberg, G., Lehtinen, H., and Friberg, S., *J. Colloid Interface Sci. 33*, **40** (1970).

19. Shinoda, K., and Kunieda, H., *J. Colloid Interface Sci. 42*, 381 (1973).

20. Rance, D., and Friberg, S., To be published.

21. Kertes, A. S., Jernstrom, B., and Friberg, S., *J. Colloid Interface Sci. 52*, 122 (1975).

22. Ahmad, S. I., Shinoda, K., and Friberg, S., *J. Colloid Interface Sci. 47*, 32 (1974).

23. Friberg, S., Mandell, L., and Ekwall, P., *Kolloid Z.u.Z. Polymere 233*, 955 (1969).

24. Soderlund, G., and Friberg, S., *Zeitschrift fur Physikalische Chemie Neue Folge 70*, 39 (1970).

APPENDIX A

MAKING PHASE EQUILIBRIA DIAGRAMS

LEON M. PRINCE

Consulting Surface Chemist
7 Plymouth Road
Westfield, New Jersey 07090

For three-component systems such as oil, water, and surfactant or water plus a surfactant and cosurfactant, a simple
rectangular graph cannot reveal as much information about the
systems as can a ternary diagram. This can take the form of a
triangular graph having three axes, one for each of the components. To simplify the drawing of these graphs, triangular
graph paper is commercially available, ruled and marked for
convenience. The virtue of the triangular graph for our purposes is that the percentage composition of the components at
any point adds up to 100%. This enables one to assign physical attributes to each point and so to visually trace changes
in these attributes with composition.

These phase equilibria diagrams are based on the Phase
Rule of J. Willard Gibbs, who promulgated it in the years
1874-1878. It was slow to be adopted as a basic tool of Physical Chemistry, but once it became understood, it was, and
is, widely utilized in the classification of heterogeneous
equilibria. It has the advantage that it contains no assumptions based on theory and is therefore immune to any changes
in our molecular or kinetic views on a given subject. Under
such circumstances, it is an ideal tool for studying micellar
solutions or microemulsions.

Where surface energy is a vital factor in the formation
and stability of the systems being studied, however, the application of the phase rule required great care. It is always
essential that the system be in true equilibrium. It is postulated that such an equilibrium may exist in any system under
given conditions when the parts of the system fail to undergo
change with time provided that the parts of the system have
the same properties when the same conditions are arrived at by
a different procedure. It is also fundamental to our situation that the phases be liquid. Since not all liquid phases

147

are soluble in one another, phase separations can occur. Moreover, being colloidal systems, mesomorphic or liquid crystalline phases may also appear to complicate the system.

In the reading of these phase diagrams, a convenient device is the use of a tie line. Such a line delineates compositions which separate into two or more phases. For example, if liquid components A, B, and C are mixed so that their total proportion indicates a composition of P, and if P lies in an unstable (immiscible) region, then the two phases which separate out will have compositions Q and R. The line connecting these points is a tie line. All compositions of the total system lying on this tie line will separate into phases having compositions Q and R; only the total amount of each phase produced will vary. Furthermore, if, when extended, a tie line passes through an apex of the triangle, corresponding to 100% of one of the components, the ratio of the other two components is the same in each phase.

One of the first to use these phase equilibria diagrams to investigate the behavior of oil, water, and surfactant systems was Per K. Ekwall of the Laboratory for Surface Chemistry of the Royal Swedish Academy of Engineering Sciences. His method of making these diagrams was described in *Acta Polytechnica Scandinavica Chem. Met. Ser. 74, I*, pp. 1-116 (1968). An excellent example of the application of his methods can be found in a paper by Ekwall, Mandell, and Fontell, *J. Colloid Interface Sci. 33*, 215 (1970). This represents very comprehensive and painstaking work. But it is of interest that his methods apply only to ionic systems which are able to withstand centrifugation.

For nonionic systems there is no method that will work in all cases. The best technique is to titrate for the liquid phases and check the boundary lines by storing samples taken from each side of the border for three months. To avoid bacterial degradation 500 ppm NaN_3 should be included in the system.

The parts of the diagram involving liquid crystals require a little more attention. Centrifugation for as short a time and as high a G value as possible is recommended. Thereafter, three months storage of samples on either side of the boundary line is advised to establish the existence of true equilibrium. Liquid crystalline phases should also be checked by optical microscopy for inclusion of solutions. Low angle x-ray scattering may also give information regarding the position of the boundary lines from discontinuities of the distance (composition) derivative. This, however, is not always a reliable technique.

Microemulsions and Tertiary Oil Recovery

VINOD K. BANSAL AND DINESH O. SHAH

*Departments of Chemical Engineering
and Anesthesiology
University of Florida
Gainesville, Florida 32611*

I. INTRODUCTION

Among various techniques suggested for tertiary oil recovery, the microemulsion flooding has received considerable
attention in recent years. Microemulsion flooding can be
applied over a wide range of reservoir conditions (1). The

use of microemulsions for oil recovery is not a recent development in petroleum technology. In 1959 Holm and Bernard (2)
filed for a patent in which the use of surfactant dissolved in
low-viscosity hydrocarbon solvent was proposed. In 1962
Gogarty and Olson (3) filed a patent describing the use of
microemulsions in a new miscible-type recovery process known
as Maraflood. In the late sixties more patents were issued
to Jones, Cooke and Holm involving microemulsions for improved
oil recovery (4). Recently Gogarty has reviewed the status
and current appraisal of the microemulsion flooding process
(5).

Generally speaking, wherever a water flood has been successful, microemulsion flooding will probably be applicable,
and in many cases where water flooding has failed because of
poor mobility relationships, microemulsion flooding can be
successful because of the required mobility control. The recovery of oil from a reservoir is basically accomplished in
three stages. In the primary oil recovery process, oil is
recovered due to the pressure of natural gases which force
the oil out through production wells. When this pressure is
reduced to a point where it is no longer capable of pushing
the oil out, water is injected to build up the necessary pressure to force the oil out. This is generally called the secondary oil recovery or water flooding process. The average
oil recovery during the primary and secondary stages is about
30% of oil-in-place. To recover at least a part of the remaining 70% oil is the purpose of the tertiary oil recovery
process. Various techniques used at this stage include carbon dioxide injection, steam flooding (thermal recovery) and
surfactant flooding by either micellar or microemulsion solutions. The microemulsion flooding technique involves a decrease of capillary forces on oil droplets in the reservoir,
thus improving oil recovery.

The microemulsion flooding process is a miscible-type
displacement process. Two basic well configurations--the
"five spot" pattern or the "line drive" pattern--are used for
the microemulsion flooding process. In the "five spot" pattern (Fig. 1), four production wells are drilled at the corners of a square, and the injection well, through which the
microemulsion is pumped, is at the center of this square. In
the "line drive" pattern, production and injection wells are
drilled in alternate rows.

In the microemulsion flooding process, the microemulsion
slug is injected into the reservoir and is followed by a
polymer solution for mobility control. This is in turn followed by the injection of water (Fig. 2). Microemulsions are
optically transparent isotropic oil-water dispersions which
can be formed spontaneously by using a combination of emulsifiers. The microemulsions used for oil recovery are composed

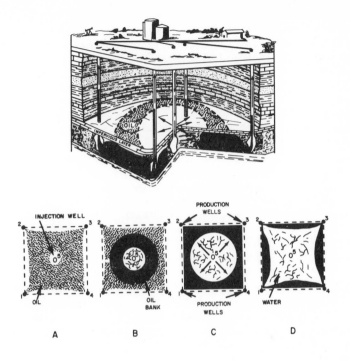

Fig. 1. The five-spot pattern of oil wells for displacement of oil in reservoirs.

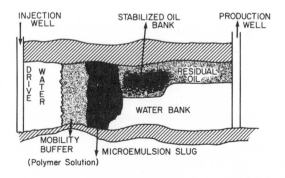

Fig. 2. A schematic presentation of microemulsion flooding process.

of hydrocarbons, surfactants, water and other organic liquids (alcohols) and are generally miscible with the reservoir oil and water. Microemulsions employed in this process may be either oil external (also called soluble oil) or water external. In most cases they contain crude oil from the reservoir in which they are to be injected.

The design of a microemulsion for a specific reservoir is basically a trial and error procedure. The formulation of the microemulsion slug for a particular reservoir depends upon the reservoir condition after the secondary recovery process and the properties of the microemulsion slug itself.

Petroleum sulfonates are the most widely used surfactants in the preparation of a microemulsion slug. The chemistry of these petroleum sulfonates and the interfacial properties of the system have to be fully understood for any successful microemulsion flooding process. Unfortunately our understanding of the theoretical aspect of the microemulsion flooding process is far from clear.

In section 2 of this chapter, the role of the capillary and hydrostatic forces on the entrapment of oil in the reservoir, and the necessary conditions for the displacement of this entrapped oil will be discussed. Following this, sections 3 and 4 will deal with important properties of a microemulsion slug required for the tertiary oil recovery. Finally, the economic aspect of the process (profitable microemulsion slug) will be discussed.

II. ROLE OF CAPILLARY AND VISCOUS FORCES ON OIL RECOVERY

Under ordinary flooding conditions (water or immiscible fluid), surface forces (capillary forces) dominate the macroscopic displacement process and are responsible for trapping a large portion of the oil within the pore structure of the reservoir rocks. The microscopic distribution of the trapped oil depends upon the hydrostatic equilibrium condition and is a function of factors such as wettability of the rock and pressure in the fluid phases. If the flood rate is made sufficiently high, however, the viscous forces dominate the macroscopic displacement process (6). In order to determine whether viscous or capillary forces are dominating the displacement process, it is convenient to consider the dependence of the displacement efficiency on a suitable dimensionless parameter. Such a number is the capillary number (N_{ca}) which is defined as

$$N_{ca} = \frac{\mu_w U_w}{\phi \gamma_{ow}}$$

(1)

where μ_W and U_W are the aqueous phase viscosity and flow rate
per unit cross sectional area, γ_{OW} is the interfacial tension
between oil and water and ϕ is the porosity of the reservoir
rock structure (7). Physically, the capillary number represents the ratio of viscous to capillary forces. The capillary
number for an ordinary waterflooding process is of the order
of 10^{-6} (7).

Laboratory studies, using either sandpacks or Berea
cores, to determine the relationship between the capillary
number and the percent residual oil saturation have been carried out by various workers (6,7). The studies involved increasing the flood rate, and/or decreasing the interfacial
tension. Sometimes an increased aqueous phase viscosity, μ_W,
was also used. All these studies were preceded by a conventional water flooding. As the capillary number was increased
by adjusting μ_W, U_W, and γ_{OW} the displacement efficiency also
increased. It appears that in order to reduce the value of
the residual oil saturation by a factor of about one half, it
is necessary to increase the capillary number (N_{Ca}) by a factor of 1000. Work reported by Foster (Fig. 3) indicates that
increasing N_{Ca} by a factor of 10,000 will result in a microscopic displacement efficiency approaching 100% (7). The displacement efficiency is basically the percentage of the oil
recovered by this method. The upper critical value of the
capillary number for 100% efficiency was found to be of the
order of 10^{-2} to 10^{-1}. The correlation between displacement
efficiency (percent residual oil) and capillary number, as
obtained from laboratory experiments under tertiary recovery
conditions, strongly suggests that the process of mobilizing
residual oil depends on a competition between viscous and
capillary forces.

Figure 4 illustrates the interplay of capillary and viscous forces in the water flooding process. Shown in the figure is water displacing oil in two capillaries of radii r_1
and r_2, respectively.

The relationship between the velocity ratio of these two
interfaces in the pore structure, and the viscous and capillary forces is given below (8):

$$\overline{V} = \frac{V_1}{V_2} = \frac{\dfrac{4Lq\mu}{\pi r_2^2 \sigma \cos\theta} + r_2^2 \left(\dfrac{1}{r_1} - \dfrac{1}{r_2}\right)}{\dfrac{4Lq\mu}{\pi r_1^2 \sigma \cos\theta} - r_1^2 \left(\dfrac{1}{r_1} - \dfrac{1}{r_2}\right)} \qquad (2)$$

where V_1 and V_2 are the velocities of the interface in capillaries 1 and 2 of radii r_1 and r_2, respectively; q is the fluid (aqueous phase) flow rate; μ is the viscosity of the fluid, L is the distance over which the capillary and viscous forces are competing (length of pore); σ is the interfacial tension between oil and water; and θ is the contact angle measured in the displacing phase.

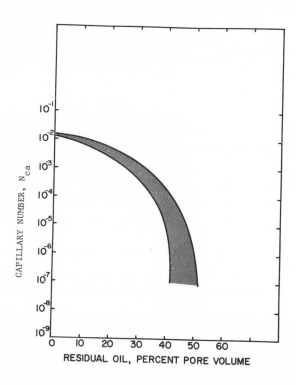

Fig. 3. The dependence of residual oil saturation of capillary number. Reproduced from Foster (7), courtesy of Society of Petroleum Engineers of A.I.M.E.

When capillary forces are negligible in comparison to the viscous forces, the last term in both the numerator and denominator in the above equation may be neglected. In such a case, the equation reduces to:

$$V = \frac{r_1^2}{r_2^2} \tag{3}$$

Under these conditions the rate of flow is proportional to the square of the radius and consequently the residual oil is left in the smaller openings.

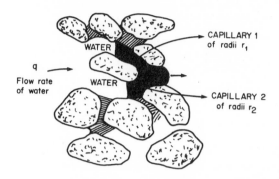

Fig. 4. A model of capillaries in a porous medium. Reproduced by Moor and Slobod (8), courtesy of Producers Publishing Co., Inc.

When the capillary forces are large compared to the viscous forces, the first term in both numerator and denominator in the equation (2) may be neglected. In such a case, equation (2) reduces to

$$\overline{V} = -\frac{r_2^2}{r_1^2} \qquad (4)$$

This suggests that the flow will be faster in the smaller capillaries and the residual oil will be left in the larger capillaries.

Taber *et al.* (9) correlated the oil displacement with the ratio $\Delta P/L\sigma$, where ΔP is the pressure drop across the distance L and σ is the interfacial tension between the oil and water. The critical ratio of $\Delta P/L\sigma$ is defined as that value below which no residual oil is produced from any of the porous media. After the critical pressure gradient was exceeded for each sample, it was possible to produce larger quantities of the residual oil by merely increasing the value of $\Delta P/L\sigma$.

Taber *et al.* (9) have shown a correlation between the $(\Delta P/L\sigma)$ critical and the viscosity of aqueous phase and oil and the permeability of porous media. The value of the critical $\Delta P/L\sigma$ ratio increases with either an increase in viscosity of the oil or aqueous phase or with a decrease in the permeability of the porous media. Their results for the residual

oil displacement from rock samples of different permeabilities
appear to have much more relevance to oil recovery than the
modest viscosity effects noted with either the aqueous or oil
fluids. The critical values of $\Delta P/L\sigma$ ranged from a high value
of 23.28 for a core with an air permeability of 95 md to a low
value of 0.31 for the most permeable sample (2190 md). The
correlation between permeability and the critical displacement
ratio $\Delta P/L\sigma$ is shown in Fig. 5. The lower permeability values

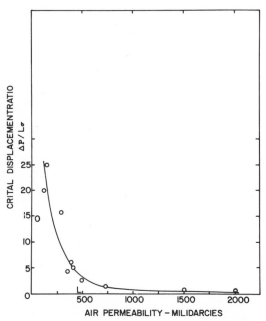

*Fig. 5. The relationship between permeability and the
critical value of $\Delta P/L\sigma$. Taber, Kirby and Schroeder (9),
courtesy of the American Institute of Chemical Engineers.*

mean that significantly higher values of $\Delta P/L\sigma$ must be
achieved before any residual oil can be displaced from a porous rock.

Stegemeier (10) has presented an "alternate path" theory
for residual oil saturation and petrophysical properties over
a broad range of varying interfacial and viscous conditions.
His model allows for intermediate water wetting up to 90°
(contact angle) and discontinuous trapped oil in single or in
interconnected pores. A quantitative expression relating the
capillary pressure to various properties of fluid and rock
was derived by Stegemeier. It was concluded that for a given
combination of fluid properties and applied pressure differential, all of the non-wetting phase will be removed from pores

having capillary pressures less than that calculated from his equation.

Slattery and Oh (11) carried out theoretical analysis for the critical pressure gradient on an ideal system of pores where the pore radius is a sinusoidal function of axial position. He concluded that for the most efficient displacement of residual oil, the porous structure should be water-wet and that intermediate wettability may be less desirable than either oil wet or water wet behavior. His estimate of the critical value of the pressure gradient agrees with the experimental data of Taber *et al.* (9) to within 50%.

III. DESIRABLE PHYSICO-CHEMICAL PROPERTIES OF MICROEMULSION SLUG

The design of a microemulsion slug for the tertiary oil recovery process is basically a trial and error procedure. However, there are some basic properties of a microemulsion slug and the effect of some particular variables on these properties that have to be studied before any laboratory and field test of a particular microemulsion slug is warranted. The success of the microemulsion flooding process for tertiary oil recovery depends on a proper choice of the chemicals that go into the formation of the microemulsion slug. The composition of the microemulsion slug is dependent upon the properties required of the microemulsion slug as well as on the conditions prevalent in the reservoir. In this section some important properties of microemulsions, as applicable to tertiary oil recovery, will be discussed.

A. Phase-equilibrium and Solubilization

Microemulsions used for improved oil recovery contain at least three components: oil, surfactant, and brine. Hence, the compositional state of the system must be specified by at least three numbers. It is, therefore, both convenient and instructive to employ a ternary representation for a phase equilibrium study.

Several studies (12-18) on ternary diagrams of microemulsion systems having application to oil recovery have been done. Using measurements of viscosity, electrical resistivity, optical birefringence and a phase disappearance technique, these studies have investigated structural changes in the microemulsion as a function of composition. A simple ternary diagram for such a three component system is shown in Fig. 6, where S, W, O and M represent surfactant, water, oil and microemulsion, respectively.

The effectiveness of the microemulsion flooding process

depends upon the extent of the single phase region in the ternary diagram of interest. The microemulsion flooding process can be prolonged by minimizing the vertical extent of the multiphase region in the ternary diagram. Reed *et al.* (15) introduced the concept of optimal salinity for a microemulsion

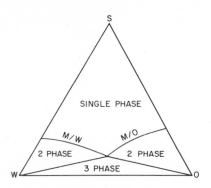

Fig. 6. Idealized ternary diagram for surfactant-oil-water systems. Reproduced from Robbins (12), courtesy of the Society of the Petroleum Engineers of A.I.M.E.

single phase to determine the minimum area of the multiphase region. As defined by them, the optimal salinity can be determined by plotting the surfactant concentration required to make a 50:50 water:oil mixture a single phase as a function of salt concentration as shown in Fig. 7. For this case, the graph exhibited a minimum near 1.25% NaCl and this salt concentration was defined as the "optimal salinity" for the given system. From the ternary diagram of the same system it was found that the minimum area of the multiphase region also occurred at 1.25% NaCl. Thus the determination of the optimal salinity for a given system assists in the determination of the formulation of a microemulsion for the tertiary oil recovery process. The effect of divalent ions (Ca^{++}) on the optimal salinity was also investigated by Reed *et al.* **(15)** (Fig. 8). They observed that Ca^{++} reduced the optimum salinity from 1.25% NaCl to 1.1% total solid (1.0% NaCl + 0.1% $CaCl_2$). These microemulsions have a salt tolerance limit beyond which they are not stable. To increase the salt tolerance of these microemulsions various alcohols (cosolvent) are added. The salt tolerance of a microemulsion is an important consideration, since the natural salt concentration in the reservoir can be very high.

The effect of various alcohols on the solubilization of brine has been studied by Jones and Dreher (19). It was observed that water soluble alcohols solubilize additional brine

into a microemulsion. Water insoluble alcohols, however,
cause solubilization of hydrocarbon, while decreasing the
brine solubility. It was also observed that for a stable microemulsion, an increase in electrolyte concentration decreases the water-insoluble alcohol requirements and reduces
the breadth of the single phase region. Conversely, an increase in electrolyte concentration increases the water-
soluble alcohol requirements.

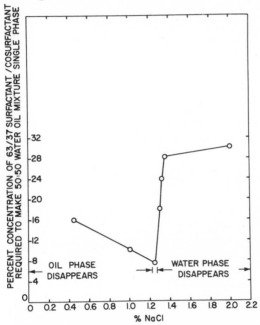

Fig. 7. The optimum salinity for mixtures of oil
and water. Reproduced from Healy and Reed (15),
courtesy of the Society of Petroleum Engineers of
A.I.M.E.

B. Phase Volume Ratio and Bulk Rheological Properties

As was discussed in the previous section the important
region of the ternary diagram for the tertiary oil recovery
process is the single phase region. In this single phase region, the microemulsion goes through a number of structural
changes between the water external and oil external extremes.
In a recent study Shah et al. (20) have shown, using electrical, birefringence and high resolution NMR measurements, that
upon increasing the amount of water in an oil external microemulsion, the structure of the microemulsion passes through
transitions from water spheres (oil external) to water cylinders to water lamellae and finally to a continuous water

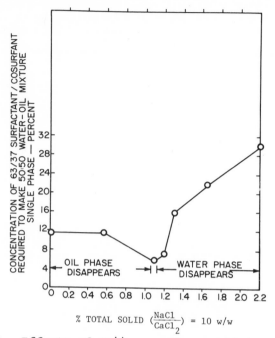

Fig. 8. *Effects of Ca⁺⁺ on optimum salinity for*
mixtures of oil and water. Reproduced from Healy
and Reed (15) , courtesy of the Society of Petroleum
Engineers of A.I.M.E.

external phase. (The figure was shown previously in Chapter
1.)   The addition of water to an oil external microemulsion
causes an increase in the size and number of dispersed water
droplets. Since the total amount of surface active agent in
the system remains constant, further addition of water would
increase the interfacial area of the water spheres, decrease
the surface concentration of the surfactant, and hence increase the interfacial tension. At some concentration of
water the intermolecular forces at the oil-water interface
would be insufficient to hold the interface together, and the
spherical droplets of water would collapse to form cylinders
of water. Using the same reasoning, further dilution with
water will cause a transition from the cylindrical structure
to a lamellar structure. Upon still further addition of water
the lamellar structure changes to spherical oil droplets which
are dispersed in a water phase.

The effect of interfacial forces and the structural
changes discussed above on the bulk viscosity of microemulsions is shown in Fig. 9 (21). The system used for relative
viscosity measurements consisted of Hexadecane (oil) + Hexanol
(cosolvent) + K-oleate (surfactant). The microemulsion used

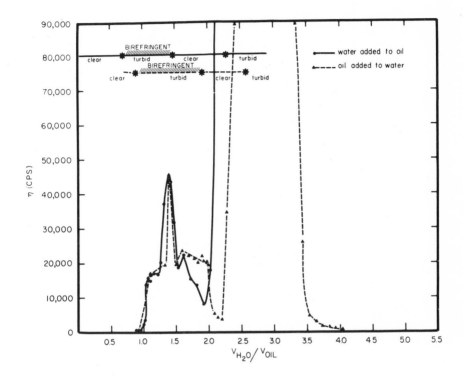

Fig. 9. The effect of water-oil ratio on viscosity of microemulsions and phase-inversion region of Hexa-decane-water-oleate-hexanol system. Reproduced from Shah, Falco and Walker (21), courtesy of American Institute of Chemical Engineers.

had a constant ratio of 10 ml hexadecane: 4 ml hexanol: 2 gm of K-oleate. The relative viscosity data for different water/oil ratios is shown in Fig. 9. From this figure it is seen that there is a maximum in the relative viscosity at a water/oil ratio of 1.4. This maximum in the relative viscosity corresponds to a lamellar structure occurring at this ratio. The viscosity peak observed between the water/oil ratios of 2.0 and 3.5 was beyond the experimentally measurable limit. It was found that both these viscosity peaks were observed upon either increasing or decreasing the water to oil (Hexadecane) ratio. It was also observed that the dispersion having a lamellar structure at the water:oil ratio of 1.4 was very viscous upon its formation, with the viscosity subsequently decreasing with time.

The effect of shearing time, at a constant shear rate, on the viscosity of these lamellar structures is shown in Fig. 10. The viscosity initially increases with shearing time and then levels off to a steady value. The effect of age of the solution on the viscosity shows that as the age increases the viscosity of these lamellar structures decreases. The initial increase in the viscosity with shearing time can be attributed to disordering and entanglement of the lamellar structure upon shearing (Fig. 11).

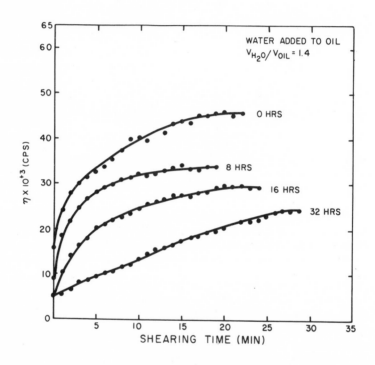

Fig. 10. The effect of shearing time on viscosity of microemulsions. Reproduced from Shah, Falco and Walker (21), courtesy of American Institute of Chemical Engineers.

In tertiary oil recovery by microemulsion flooding, the stability of the microemulsion slug is an important considera-tion. In particular its mobility (as defined in a later sec-tion) has to be controlled. The mobility is inversely propor-tional to its viscosity and hence a very large viscosity is undesirable. This would cause plugging of the pores with con-sequent loss of surfactant. Hence it is advisable to stay out of the range of gel formation (lamellar structure) or to avoid

it by adjusting the composition of the microemulsion. Alternatively, a shift in the viscosity peak could be achieved
through a manipulation of the salt, oil, surfactant or cosolvent concentrations.

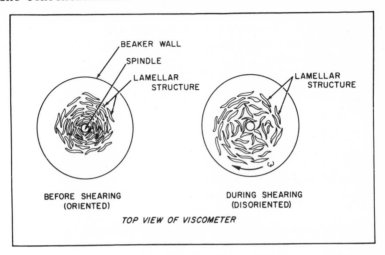

*Fig. 11. The effect of shear time on the orientation
of lamellae in a lamellar liquid-crystalline phase.
Reproduced from Shah, Falco and Walker (21), courtesy
of American Institute of Chemical Engineers.*

C. Interfacial Tension

The interfacial tension between the crude oil and the
displacing liquid (microemulsion) is one of the major parameters which has to be controlled and fully determined before
the microemulsion slug can be used for any tertiary oil recovery process. Healy and Reed (22) used the Laplace equation

$$\Delta P = 2\gamma \ (\frac{1}{r_1} - \frac{1}{r_2}) \tag{5}$$

to calculate the pressure difference (ΔP) across an oil drop
having a curved interface with principal radii r_1 and r_2, and
an interfacial tension γ between the oil and water. Using an
electroscan micrograph of a cross section of the porous medium
they calculated the pressure the water would have to develop
in order to displace the largest entrapped oil droplet observ-
able in the micrograph. This would correspond to the lowest
pressure drop (largest drop size) necessary to begin displacing the oil. Using a normal oil/water interfacial tension of

the order of 10 dynes/cm, the calculated pressure drop turned
out to be $\approx$ 5 X 10^2 psi/ft, while a practical limit to pressure drops achievable under field conditions is about 1-2 psi/
ft. Under this practical limitation, the most obvious solution to displacement of the entrapped oil is through a reduction in the crude oil/displacing liquid interfacial tension to
about 0.001 dyne/cm.

The microemulsion slug used in the tertiary oil recovery
process should effectively displace oil at the front, and
should be effectively displaced by drive water at the back, as
illustrated in Fig. 2. For an efficient process, both these
are essential requirements, and therefore the interfacial tensions at the microemulsion-oil interface (γ_{mo}) and the micro-
emulsion-buffer solution interface (γ_{mw}) should be very low as
has already been discussed. The conditions where these two
interfacial tensions are both low and equal to each other is
of particular significance in designing a microemulsion slug
for the process. The equality condition arises from the definition of the capillary number and helps to ensure stable
movement of various banks. A large difference in N_{Ca} at the
two interfaces would lead to different pressure gradient requirements for each interface.

Healy and Reed (22) have studied the effect of NaCl concentration on the interfacial tension γ_{mo} and γ_{mw} and the effect on the solubilization parameters V_o/V_S and V_w/V_S (where
V_S = volume of surfactant in the microemulsion not including
cosolvent, and V_o and V_w are the volumes of oil and water in
the microemulsion phase, respectively). The results obtained
by them are shown in Fig. 12. It was found that an increase
in salinity decreases γ_{mo} and increases γ_{mw}. The point of
intersection of γ_{mo} and γ_{mw} was defined as the "interfacial
tension optimal salinity" (C_γ). Determination of the interfacial tension optimal salinity assists us in designing a microemulsion slug because of the above stated conditions for
γ_{mo} and γ_{mw}. The phase behavior optimal salinity, C_ϕ, is defined by the intersection of V_o/V_S with V_w/V_S. The correlation between C_γ and C_ϕ is apparent, and this indicates that
phase volumes can replace interfacial tension measurements as
a preliminary measure of interfacial activity, thus markedly
reducing the labor involved.

The effect of hexanol (cosolvent) and K-oleate (surfactant) on the interfacial tension was determined by measuring
the average drop volume of water in hexadecane by Shah (20).
This method consists of squeezing the smallest possible water
drop out of a microsyringe into a bath of hexadecane. The
smaller the drop volume of water, the lower is the interfacial
tension. The results of these measurements are given in Table
I. This table illustrates the effect of the various constitu-
ents in a microemulsion (Hexadecane + Hexanol + K-oleate +
water) on the average drop volume of water (with or without

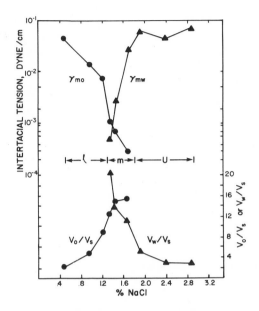

Fig. 12. The effect of salt concentration on the interfacial tension and solubilization parameter in oil-brine-surfactant-alcohol system. Reproduced from Healy and Reed (22), courtesy of American Institute of Chemical Engineers.

additives) in hexadecane (with or without additives). The re-sults show that Potassium oleate, even at low concentrations, is more effective than hexanol in decreasing the average drop-volume of the water drops. However, the presence of hexanol and potassium oleate in hexadecane and water, respectively, decreases the drop volume to less than 10^{-6} ml. This decrease in drop volume by the addition of both hexanol and K-oleate indicates the drastic decrease in interfacial tension between Hexadecane and water made possible by this addition.

In practice various petroleum sulfonates have been em-ployed in the formulation of microemulsions which exhibit ultra-low interfacial tension with oil as well as with the aqueous phases. It has been shown (23) that a petroleum sul-fonate with an equivalent weight distribution that is rela-tively narrow and/or symmetrical about the median is the most effective in lowering interfacial tensions. A minimum in in-terfacial tension can also be achieved through an adjustment of the electrolyte content of the aqueous phase (23). Sodium chloride was found to be more effective than sodium sulfate, carbonate or tripolyphosphate in decreasing the interfacial tensions between the oil and the aqueous phase (23).

TABLE I

Effect of Hexanol and Potassium Oleate on the Average
Volume of Water Drops in Hexadecane

Interfacial Composition		Average Drop-Volume
Hexadecane (oil)	Water	
No additive	No additive	2.8×10^{-2} ml
Hexanol		
0.1 ml/ml of oil	"	6.0×10^{-3} ml
0.4 ml/ml of oil	"	5.0×10^{-3} ml
0.6 ml/ml of oil	"	5.0×10^{-3} ml
No additive	K-oleate	
"	0.02 gm/ml of water	8.0×10^{-4} ml
"	0.56 gm/ml of water	8.0×10^{-4} ml
"	0.1 gm/ml of water	4.4×10^{-4} ml
Hexanol	K-oleate	
0.025 ml/ml of oil	0.06 gm/ml of water	1×10^{-4} ml
0.05 ml/ml of oil	"	6.6×10^{-5} ml
0.075 ml/ml of oil	"	volume $< 10^{-6}$ ml

IV. MOBILITY CONTROL DESIGN FOR THE MICROEMULSION PROCESS

Adequate mobility control is an important requirement for
any microemulsion flooding process. Mobility control of the
process implies changing the properties of the injected fluids
such that a stable movement of the separate banks is achieved
with a minimum of mixing and dispersion. The mobility of a
fluid is defined by $\lambda = k/\mu$, where λ is the relative mobility
of the fluid with a relative permeability of k and a viscosity
μ. Without a mobility control slug following the microemulsion slug, the integrity of the microemulsion slug is jeopard-
ized after its injection into the reservoir, and large pore
volumes of the microemulsion slug are required for the process
to proceed. Mobility control for the process is based upon
the total mobility of the oil and water flowing ahead of the
slug in a stabilized bank. For an ideal system, the mobility
of the microemulsion slug is made equal to or less than that

of the oil bank. The mobility of the buffer is made equal to
or less than that of the microemulsion slug.

The mobility of a stabilized oil bank is calculated (24)
using the following equation

$$q_{tb} = KA \ (\frac{(\Delta P)}{(\Delta L)})_b (\lambda_{rw} + \lambda_{ro})_b \qquad\qquad (6)$$

where q_{tb} = Total flow rate in the stabilized oil/water bank

K = Absolute permeability

A = Cross sectional area

$\frac{(\Delta P)}{(\Delta L)_b}$ = Average pressure gradient in the stabilized
oil/water bank

$(\lambda_{rw} + \lambda_{ro})_b = [\frac{K_{rw}}{\mu_w} + \frac{K_{ro}}{\mu_o}]_b$ = Total relative mobility in
the stabilized oil bank

where K_{rw} = Relative permeability with respect to water

K_{ro} = Relative permeability with respect to oil

and μ_w and μ_o are the viscosity of water and oil, respectively.

The total relative mobility can be obtained as a function
of water saturation from relative permeability data for the
aqueous phase and the oil phases (24). The use of such data
enables the appropriate fluid viscosity and total relative
mobility to be plotted as a function of water saturation. A
typical curve for a reservoir plotting the total relative
mobility as a function of water saturation is shown in Fig.
13. The stabilized oil bank should not have a mobility less
than the minimum seen in the figure, and thus this minimum
mobility value represents a design mobility value.

A. Microemulsion Slug Mobility

The first step in a mobility design procedure is the control of the mobility of the microemulsion slug. The mobility
of a microemulsion is a function of its composition and can be
controlled to fit a specific application. The parameters that
can be varied to control the mobility of the microemulsion
slug are the amount of water, the electrolyte concentration,
the type of hydrocarbon and surfactant used and the use of cosurfactants. Care should be taken not to make a change in
mobility control that adversely affects the other properties
of the microemulsion and consequently its ability to displace
the oil.

Figure 14 (25) illustrates the effect of water content on
the viscosity of a microemulsion. The microemulsion contained
a constant ratio of 76% Pentane to 19% sulfonate to 5% Isopropanol. Pentane, sulfonate and Isopropanol together represent

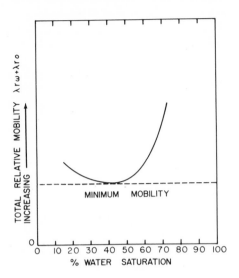

*Fig. 13. The effect of percent water saturation on
the total relative mobility of oil and water. Reproduced from Gogarty, Meabon and Milton (24), courtesy
of the Society of Petroleum Institute of A.I.M.E.*

one component and the water content (the second component) was
varied to obtain different compositions. Upon addition of
water to an oil external microemulsion there is an exponential
increase in the microemulsion viscosity up to a water concentration of about 45%. This is the inversion point for this
system and represents the point where the oil-external character of the system switches over to a water external one. Figure 14 also illustrates the use of electrolyte in controlling
the viscosity. It was found that the viscosity could be reduced by up to two orders of magnitude with the addition of
less than 0.3% Na_2SO_4 to the above microemulsion. It was also
observed that the addition of cosurfactant caused a reduction
in viscosity. Usually an order of magnitude reduction in viscosity is not uncommon with the addition of 1 to 2 percent of
either isopropanol or normal butyl alcohol.

B. Mobility of Buffer (Polymer) Slug

For an efficient microemulsion flooding process the mobility of the buffer displacing the microemulsion slug is one of
the important factors to be taken into account in designing
the process. As mentioned, for a stable system, the mobility
of the buffer solution must be equal to or less than the mobility of the microemulsion slug (24). Higher mobility of

buffer solution causes the "fingering" of polymer solution
into the microemulsion slug. Water thickened by the addition
of a polymer serves as an effective mobility buffer solution.
Many polymers have been reported to be effective mobility
control agents (26). However, polyacrylamides are the only
polymers that have been used as mobility control agents on a
large scale. Mobility control with polyacrylamides is
achieved through a reduction in both the viscosity and the permeability (27). Both the molecular weight and the degree of
hydrolysis of the polymer are important characteristics in the
design of a mobility control solution (27).

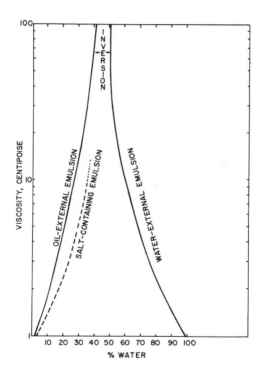

*Fig. 14. The effect of percent water on the viscosity
of microemulsions. Reproduced from Gogarty and Tosch
(25), courtesy of the Society of Petroleum Institute
of A.I.M.E.*

V.   ECONOMIC ASPECTS OF THE PROCESS

A.   Economic Aspects of Mobility Control

The cost of the mobility buffer solution and microemulsion
slug depends upon their respective compositions. The

microemulsion slug mobility can be reduced to almost any value
by changing the composition of the microemulsion, and normally
the cost variation entailed in changing the composition of the
microemulsion is insignificant. The mobility of the microemulsion slug can thus be adjusted to a value much lower than
the design mobility without significantly affecting its cost.
However, a low value for the mobility of the microemulsion
slug requires a higher concentration of polymer (in the buffer
solution) in order to ensure adequate mobility control at the
buffer-microemulsion slug interface. From an economic standpoint, the cost of the mobility buffer solution is related to
its polymer concentration, and hence, the polymer concentration should be kept to a minimum (1). The cost of the polymer
used in the mobility buffer imposes a restraint upon the lowering of the slug mobility. For this reason the mobility of
the buffer solution is made only slightly less than the design
mobility. Cost of buffer permitting, buffer injection con-
tinues until the flood is completed. This is because the injection of drive water behind the mobility buffer can cause an
unfavorable condition, since the drive water can penetrate and
bypass the mobility buffer and slug and change the process
from tertiary to ordinary water flooding.

B. Economic Aspects of the Microemulsion Slug

The cost of the chemicals used in the formulation of microemulsions for the flooding process and the oil saturation
in the reservoir at the time the process is initiated determine whether the process is economically feasible or not. Of
all the chemical components making up a microemulsion slug,
the surfactant, petroleum sulfonate, is the most expensive
component. The cost of the microemulsion can be decreased
considerably by developing a new formulation that uses a lower
surfactant concentration, lower cosurfactant concentration and
also by using the crude oil in place of refined hydrocarbons.
Economic success or failure of the microemulsion flooding process depends largely upon the proper choice of a microemulsion
slug size. Jones (28) describes a simple technique for estimating "optimum slug size." Optimum slug size is defined as
that slug size that will maximize the profit. The data required to determine the "optimum slug size" includes oil price,
average oil saturation, per-barrel slug cost, and a slug size
vs. oil recovery curve. The slug size versus oil recovery
curve can be obtained either from laboratory tests or pilot
field tests, preferably from the latter. According to his
derivation maximum profit would occur when the following condition is satisfied:

$$\frac{\delta R_O}{\delta V_S} = \frac{C_S}{S_O P_O} \tag{7}$$

where $C_S/S_O P_O$ represents the slope of the oil recovery vs.
slug size curve
and R_O = oil recovery as fraction of oil in place before
the flood
   V_S = slug volume, fraction of reservoir pore volume
   C_S = cost of injected microemulsion slug
   S_O = average oil saturation before flooding
   P_O = price received for oil after royalties.
The point where the tangent having a slope $C_S/S_O P_O$ touches the
oil recovery vs. slug size curve represents the most profitable slug size (Fig. 15). The intercept of the tangent with
the ordinate represents the net oil recovery after subtracting
slug costs. It can be seen from Fig. 15 that the greater the
slope, $C_S/S_O P_O$, the higher the slug cost and the lower the net
oil recovery. If the slope of the tangent is greater than the
slope of any portion of the oil recovery-slug size curve the
process will not be economically feasible. Though this technique is very quick and convenient to determine the optimum
slug size for an economically feasible process, other economic
factors, such as time value, sometimes make the process un-
economical.

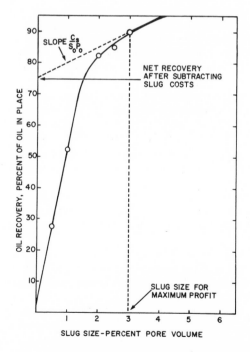

*Fig. 15. The determination of most profitable slug
size for tertiary oil recovery. Reproduced from
Jones (28), courtesy of the Society of Petroleum
Institute of A.I.M.E.*

REFERENCES

1. Poettman, F. H., Secondary and Tertiary Oil Recovery Process, Interstate Oil Compact Commission, Oklahoma, p. 82 (1974).
2. Holm, L. W., and Bernard, G. G., U.S. Patent No. 3082822 (1959).
3. Gogarty,W. B. and Olsen,R. W.,U.S.Pat. 3254714 (1962).
4. Cooke, C. E., Jr., U.S. Patent No. 3373809 (1965); Holm, L. W., U.S. Patent No. 3482632; Jones, S. C., U. S. Patent No. 3497006 and 3506070.
5. Gogarty, W. B., 81st National Meeting of AIChE, Kansas City, Missouri (1976).
6. Taber, J. J., *SPE J. 9*, p. 3 (1969).
7. Foster, W. R., *J. Pet. Tech.*, p. 205 (Feb., 1973).
8. Moor, T. F., and Slobod, R. L., *Prod. Monthly 20*, No. 10, p. 20 (1956).
9. Taber, J. J., Kirby, J. C., and Schroeder, F. V., Paper 476, 71st National AIChE Meeting, Dallas, Texas (1972); "Declining Domestic Reserves Effect on Petroleum and Petrochemical Industry," edited by G. H. Cummings and W. B. Franklin, *AIChE Symposium Series 127, Vol. 69*, p. 55 (1973).
10. Stegemeier, G. L., SPE 4754 presented at SPE Symposium on Improved Oil Recovery, Tulsa, Oklahoma (1974).
11. Slattery, J. C., and Oh, S. G., ERDA Symposium on Enhanced Oil Gas Recovery, Tulsa, Oklahoma (1976).
12. Robbins, M. L., SPE 5839 presented at SPE Improved Oil Recovery Symposium, Tulsa, Oklahoma (1976).
13. Robbins, M. L., Paper presented at 76th National AIChE Meeting, Tulsa, Oklahoma (1974).
14. Anderson, D. R., Bidner, M. S., Davis, H. T., Manning, C. D., and Scriven, L. E., SPE 5811 presented at SPE Improved Oil Recovery Symposium, Tulsa, Oklahoma (1976).
15. Healy, R. N., and Reed, R. L., *SPE J. Vol. 14*, p. 491 (1974).
16. Healy, R. N., Reed, R. L., and Carpenter, C. W., *SPE J. Vol. 15*, p. 87 (1975).
17. Healy, R. N., Reed, R. L., and Stenmark, D. G., SPE 5565, presented at Fall SPE Meeting, Dallas, Texas,(1975).
18. Healy, R. N., and Reed, R. L., SPE 5817, presented at SPE Improved Oil Recovery Symposium, Tulsa, Oklahoma (1976).
19. Jones, S. C., and Dreher, R. D., SPE 5566, presented at Fall SPE Meeting, Dallas, Texas (1975).
20. Shah, D. O., Tamjeedi, A., Falco, J. W., and Walker, R. D., Jr., *AIChE J. Vol. 18*, p. 1116 (1972).
21. Shah, D. O., Falco, J. W. and Walker, R. D., Jr., *AIChE J. Vol. 20*, p. 510 (1974).

22. Healy, R. N., and Reed, R. L., Paper presented at
    81st AIChE Meeting, Kansas City, Missouri (April, 1976).
    1976).
23. Wilson, P. M., Murphy, C. L., and Foster, W. R.,
    SPE 5812 presented at SPE Improved Oil Recovery
    Symposium, Tulsa, Oklahoma (1976).
24. Gogarty, W. B., Meabon, H. P., and Milton, H. W., Jr.,
    J. Pet. Tech., p. 141 (1970).
25. Gogarty, W. B., and Tosch, W. C., *J. Pet. Tech.*,
    p. 1411 (Dec., 1968).
26. Trushenski, S. P. Dauben, D. L., and Parrish, D. R.,
    SPE 4582 presented at SPE Annual Fall Meeting,
    Las Vegas, Nevada (1973).
27. Gogarty, W. B., *J. Pet. Tech.*, p. 161 (June, 1967).
28. Jones, S. C., *J. Pet. Tech.*, p. 993 (1972).

Index

A

Activity, 79, 125
Alkyd emulsion, 5, 22, 31, 92, 102
AMP, *see* 2–amino–2–methyl–1–pro-
panol, 73, 74, 81, 103, 110, 113
Amphiphile, *see* Cosurfactant, 28,
76, 92, 96, 95, 105
Anesthetics, 32
Appendix, 40, 51-56
Appendix A, 147-148
Asphalt, 51

B

Bacterial degradation, 148
Beeswax, 22, 24, 46
Benzene, 92, 95, 96, 101, 102, 104,
105, 112, 117, 118, 127, 128
alcohol ratio constant of, 99
p–dimethyl, 70
Berea cores, 153
Bile Acids, 32
Bimolecular leaflets, 93, 98
Biological membranes, 123
Birefringent, optical streaming, 6-
17, 96, 98, 101, 115, 116, 118,
157, 159
Blood, artificial, 32
Borax, 25, 28, 46, 103
Boric acid, 25, 41, 103

Bowcott and Schulman Paper, 92-
93, 98-101
Brownian Movement, 14, 106

C

Candelilla wax, 22, 103
Capillary forces, 152, 153, 155
Capillary number, 152, 153, 164
Capillary pressure, 156, 157
Carnauba wax, 5, 12, 21-30, 48, 92,
101, 102, 109
Cation, 40
asymmetric disorder of, 102, 103
hydrated of, 122
larger of, 45
size of, 99
size vs. anionic tail of, 10
water sheath, 109
Centrifugation, 4, 13, 85, 148
CER, cohesive energy ratio, 38, 43
Chaotropy, 110
Chemical potential, 111, 124
Chlordane, 5, 22, 29, 30, 48, 92
Cleaning, dry, 4, 31, 51, 64, 92
waterless hand of, 32
Colognes, 32
Cloud point, 62, 74
Cosmetics, 12, 58, 120
Cosulubilized, 116, 117, 121
Cosurfactant, 3, 40

U, V, W, X

7
8
9
0
1
2
3
4
5